Project Execution of Mega-Projects for the Oil and Gas Industries

Project Execution of Mega-Projects for the Oil and Gas Industries

Soosaiya Anthreas

CRC Press is an imprint of the
Taylor & Francis Group, an informa business

First edition published 2021
by CRC Press
6000 Broken Sound Parkway NW, Suite 300, Boca Raton, FL 33487-2742

and by CRC Press
2 Park Square, Milton Park, Abingdon, Oxon, OX14 4RN

CRC Press is an imprint of Taylor & Francis Group, LLC

ISBN: 978-0-367-67525-7 (hbk)
ISBN: 978-1-003-13165-6 (ebk)

Typeset in Times
by codeMantra

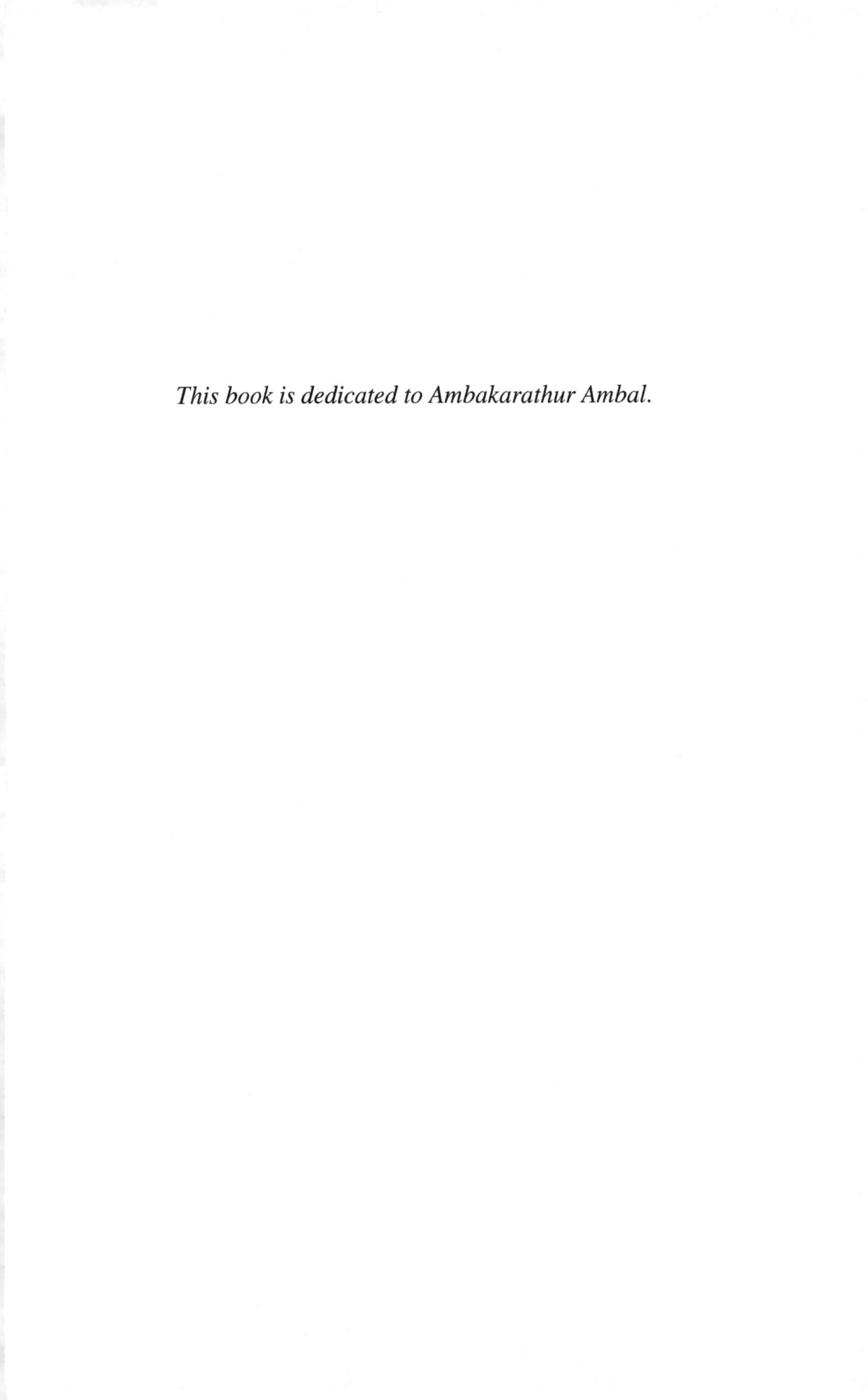

This book is dedicated to Ambakarathur Ambal.

Contents

SECTION II Commercial Aspects

SECTION III Technical Aspects

SECTION IV Glossary of Abbreviations, Names, and Meaning

Foreword

Investments in mega-hydrocarbon projects can cost several billion dollars. Investors are drawn to take advantage of the economies of scale and seek forward and backward integration. Considering the demand for hydrocarbons has been high and will continue to do so in the foreseeable future, investors are not averse to pumping in a lot of money into these projects, even borrowing huge sums for the purpose. Consequently, the pressure on completing these projects in the shortest possible time is enormous. Also, such projects are extremely complex to execute given the various variables and huge risks involved and many crucial decisions have to be taken spontaneously.

A budding project engineer can be quite overwhelmed with the pace and complexity of such projects, and to be a key player in the team, he needs to quickly size up situations and contribute proactively to the project management process.

It is for such young engineers that this book offers a simple but comprehensive oversight of project management, especially of mega-hydrocarbon projects. Having spent more than three decades in this industry and rising through the ranks to occupy top managerial positions, the author is abundantly qualified to provide the overview that a budding project engineer may seek. This book is sans academic jargon and concise, and introduces topics in such a way that with the basic understanding obtained, the project engineer can quickly acquire further skills on the job.

This book is also immensely helpful to engineers working on proposals and developing project execution plans – all of which are carefully looked into by Owner companies from large EPC Contractors, who are bidding for the job. The successful EPC Contractors lay great emphasis on the quality of the project management team and would not hesitate to pay the right price to secure them. Therefore, this book provides career progression opportunities to budding engineers.

I have known the author for more than a decade. He has successfully managed mega-EPC projects with multibillion dollar outlay. An EPC project if carefully executed can bring in rich returns, and at the same time, a flawed execution brings huge losses and could challenge the organization's very existence. Coming from such a hands-on project manager, this book will strike a chord with the young engineer and will provide the connect that the reader seeks – of personal experience.

V. Viswanathan
August 2020

Preface

The author finds an utmost necessity to provide a guide book to the professionals involved in mega-industrial project execution, especially the engineers and supervisors thereof. Besides, it can help researchers and final-year students who are interested in knowing how a project of mega-scale is executed. The project execution focuses on projects where the Contract is in Lump Sum Turnkey EPC format.

This book will be known for its unique qualities as the author, with 33 years' hands-on experience, draws the essence directly from the project execution front and from the culture of organization known for Good Industry Practice in global standards. This book thus likely differs differ from any others bearing similar titles.

This book is divided into three sections: The first addresses tasks/activities that fall under the scope of each discipline, the second commercial aspects, and the third is technical.

The material in each section is so chosen that it should fill the void that normally confronts the new engineers and supervisors while handling challenging issues in the execution of projects. Even for the experienced project engineer who is an expert in one discipline, this book will be of great help because it can give him full a picture of the project and insights across all disciplines. This book can also serve as a training tool for employees in organizations.

Section I breaks down the whole project into smaller elements (tasks and activities) in Level 3 platform minutely and completely. In other words, the sum of all these tasks constitutes the full project, ironically defeating the gestalt principle. Though the activities and tasks are drawn from the body of the oil and gas project, the pattern still serves good for other industrial projects.

Section II addresses those commercial issues which usually confront engineers, fresh or experienced, and create greater misunderstanding between Contractor and his Subcontractors/Vendors in the execution phase. Such misunderstanding spoils the relationship and leads the parties to the battlegrounds of claim and dispute. The author has endeavored to provide his insights to create confidence among engineers so that they can handle professionally and settle amicably the issues. The insights provided would also help engineers involved in the project proposal stage.

Section III addresses those technical issues which assume importance and draws attention during the execution phase.

Further in the appendix section, there is a collection of important resource materials placed under 11 appendices related to the execution of mega-projects in oil and gas industries.

Preface

Acknowledgments

My gratitude stands forever to my friends Mr. V. Viswanathan GM, HP Mittal Energy Ltd, and Mr. Perumal Kumar, Construction Manager, Technip FMC, Vizag Refinery Expansion Project Site, for their comprehensive reviews and valuable comments as experts in the field of construction of mega-industrial projects in oil and gas and petrochemical sectors.

I laud, affectionately, my daughter Ms Jzacksline Andreela Walton, Product Owner (IT), for her dedication in finding the publisher, besides providing review comments, my wife Danee Joycee, Rtd. Associate Professor (Civil Engineering), and son Andrei Crichton for their standing beside me during the writing.

Above all, to the Almighty who stands near.

Author

Soosaiya Anthreas graduated from CIT in Mechanical Engineering. He has 34 years of experience from mega-industrial projects in the field of project development, project management, and project construction. He worked in global EPC companies like Samsung Engineering and Technip FMC. As a weekend novelist, he published his first literary fiction novel, *The Dance of the Sea*. He lives in Chennai, India with wife and son.

Introduction

In Industry Practice worldwide, of several execution models of projects executed so far, the EPC execution delivery model on lump sum basis has been found to be efficient and popular.

In this book, the EPC Model of project execution is explained in Section I, commercial topics in Section II, and technical topics in Section III. The inputs for this explanation come more from experience of the author than from the theory. EPC Model is sometimes referred to as "EPCC Model".

A project consists of various units when considered in process and utility point of view, and various areas in geographical point of view as are shown in Appendices 6 and 11, respectively. An EPC Contractor's work scope is assumed to include detailed engineering, procurement, construction, pre-commissioning, commissioning, start-up/initial operations, and performance test. Work scope in brief (Level 1) is shown below.

For simplicity, the term "EPC Contractor" is hereinafter referred to as "Contractor" and "EPC Contract" as "Contract".

Detailed engineering is assumed to take place in the Home office of Contractor who provides furnished/equipped office to Owner's engineers so that closer proximity and control is exercised by Owner on Contractor's work process.

So does the procurement.

Once the 60% progress is achieved in engineering, the workstation is moved from Home office to Site office. Contractor's scope includes construction of Site office, accommodation, and other temporary facilities.

The parcel of the land (Site for construction and temporary facilities) is made available with fence and security system to Contractor right from the commencement date or zero date. Owner's obligation includes providing the feedstock and other utilities (Table 1).

TABLE 1
Work Scope in brief

No	Scope Division @ Level 1	Contractor	Owner
1	Front End Engineering Design (FEED) and Soil Data		X
2	FEED review	X	
3	Detailed engineering design	X	
4	Procurement/supply/transport	X	
5	Construction	X	
6	Pre-commissioning/commissioning/testing	X	
7	Land		X
8	Feedstock		X

Section I

Planning and Execution in Level 3

For explaining various aspects at Level 3 platform, an upstream natural gas treatment project is assumed, where the project contains a central processing facility (to be called as Main Plant), and well heads, flow lines, clusters, and trunk lines (to be called as Off-Site). The Main Plant and Off-Site together during the execution stage are called project, which on completion stage called the Plant. The project features are given in Table 1.

The activities indicated and addressed in the following chapters under Section I constitute the tasks for Level 3 Detailed Schedule. These are the tasks that Contractor discloses to Owner officially and gets them populated in Level 3 Schedule. Contractor does not disclose those deliverables (tasks) that he prepares and monitors for his in-house internal use, and hence, they do not appear in Level 3 Schedule.

Key Dates and Contract Milestones are reflected here.

Any documents that Government authorities require at the start of the project regarding project are also stated as tasks here.

TABLE 1
Project Features

Plant capacity:	6 million standard cubic meter per day from 20 wells.
Products:	Natural gas and condensates
Contract price:	1 billion USD
Timeline:	40 months
Major process units:	Slug catcher, mercury removal, compression, acid gas removal, dehydration, and due point unit, all being part of central processing facility under Main Plant.

In Section I, planning and execution is explained through ten topics (main activities): (1) Management, Control, and Administration; (2) HSE, Social, and Security; (3) Services and Facilities to Owner; (4) Engineering; (5) Procurement and Supply; (6) Transport, Logistics, and Storage; (7) Construction; (8) Pre-commissioning; (9) Commissioning, Start-up, and Performance Testing; and (10) Temporary Facilities. The breakdown aforesaid is comparable to the first ten items of Appendix 2. The last two main activities (Final Documentation and Owner Operators Training) of Appendix 2 are not treated in this book. However, the Scheduler should take these two activities into account while planning in Level 3.

1 Management, Control, and Administration

1.1 MANAGEMENT

The management assumes paramount importance in EPC Execution as EPC Contractor assumes sole and full responsibility until the Plant performs as designed at the scheduled date, including safety and quality. No projects succeeded without a good management. With good management, even difficult-to-execute projects or underquoted/bid projects have performed commendably with profits at the end. For a sustainable growth of management towards excellence, systems and procedures are important.

Normally, the management will comprise as follows, including the support from respective centers:

1. Contractor's Main Management Center (at Head office),
2. Contractor's Engineering and Procurement Center (at Head office),
3. Contractor's Construction Management Center (at Head office and Site),
4. Interface Management.

Under Level 3, collectively, the tasks involved among the above-mentioned centers include the preparation of

- Co-ordination and Communication Procedure;
- Mobilization plan and Chart;
- Temporary Facility Establishment Philosophy;
- Engineering Execution Philosophy;
- Procurement Execution Philosophy;
- Construction Execution Philosophy;
- Front End Engineering Design (FEED) and overall Design Review Program;
- Project Review Dossier (including the result on FEED verification, the final technical and commercial negotiation at the last leg of bidding process, and highlights/focus on the critical and important activities);
- PMO's Roles and Responsibilities;
- Change Procedure;
- Close-out Procedure;
- Corporate Audit Philosophy on the Project;
- Interface Management Plan.

In addition to the above deliverables, the Level 3 Schedule reflects that management, administration, and secretarial support services from Head office are available until project completion under Contractor's Main Management Center, Contractor's Engineering and Procurement Centers, Contractor's Construction Management Center, and Management from Construction Site.

Appendix 1 explains how above managements are integrated and organized. Engineering and Procurement Centre is located at the Home office of Contractor, while Management for Site is located at Site. Interface Management during engineering phase is located at Home office of Contractor, while during Construction, it is located at Site. The Project Manager is the overall head of all above management.

Contractor prepares and follows the procedures and plans. A typical list is shown in Appendix 10. The head of the project management is Project Manager who assumes full responsibility and authority for the Contract with power of attorney given by the Chief Executive Officer or competent authority of the Contractor's company.

Interface Management assumes importance when the Owner has many EPC Contractors working simultaneously in the complex where the project is executed.

Steering Committee or Sponsor Committee

The project has a steering committee of sponsors from both Owner and Contractor. The sponsors generally are top executives in the rank of directors. In a project, critical issues may arise to block the progress of the project. Such issues may sometimes remain unresolved at Project Manager's level, but can be resolved at steering committee level. Steering committee normally monitor important issues at distance daily but, however, refrain from interfering unless situation so warrant the intervention from steering committee. Steering committee has authority to remove Project Managers, approve expenses of larger sum/overruns of budgeted cost, and appoint key personnel. Any sum exceeding budgeted limits under respective subcategories shall be subject to the approval from sponsors.

Steering committee or sponsor meetings provide a platform to resolve any issues that threaten to adversely impact on the project's important milestones. On monthly basis, Project Manager submits to the respective sponsors a report informing on various blockages and issues that throw challenges to the execution team in a PowerPoint presentation on the above, which can be later used in the steering committee meeting.

If the project runs in critical path with delays and has issues of concerns, the frequency of such meetings is increased to occur once a month, otherwise once in 2 months. Steering committee members assume ultimate powers to resolve those issues which remain unresolved at Project Manager's level. However, if issues still continue to remain unresolved, then such matter is escalated to the section of dispute. When dispute is notified, parties pursue resolution as per dispute procedure (basically arbitration as provided in Contract). As arbitration involves considerable expenses, both parties make efforts to resolve amicably the issues through dedicated meetings. The appropriate time for resolution is the time at which project inches closer to achieving a major milestone. In Industry Practice, icebreaking happens when both parties come forward to share responsibility for delay or cost impacts on equal measure that is 50:50.

Project Management Organization

Appendix 1 reflects a typical organization.

At the initial phase (until 60% model review or 70% engineering), the Project Manager is located at Contractor's Home office where engineering and procurement are carried out and is supported by Project Control Manager, Project Engineering Manager, Project Procurement Manager, Quality Control Manager, and Health Safety Environment (HSE) Manager. Each of these managers has subordinates as stated below:

Project Control Manager (PCM)

PCM typically has a team of engineers as below:

- Schedule Controller;
- Cost Engineer;
- Document Controllers;
- Administrative and Reports Engineer;
- Contract Manager.

Project Engineering Manager (PEM)

PEM typically has a team of engineers as below:

- Project Engineer Process;
- Project Engineer Piping;
- Project Engineer Electrical and Telecommunications;
- Project Engineer Instrumentation;
- Project Engineer Civil, Structural, and Architecture;
- Project Engineer Rotating equipment and Packages;
- Project Engineering Static equipment;
- Project Engineer Heating Ventilation Air Conditioning (HVAC).

The core engineering group is different from project engineering group. Each discipline will have one Lead Engineer (LE). Project Engineer is the interface between Owner and Contractor in engineering matters of administrative nature. However, for resolution of long pending or urgent issues, LE, together with Project Engineer, is expected to take up this with the Owner. Project Engineer closely co-ordinates with Document Controller for issuance and distribution of engineering deliverables and keeps track of progress, Minutes of Meeting (MOM), and monitor and resolve blocking points.

PEM manages the following LEs. LEs being part of core engineering group will report to their boss in core engineering group regardless of project period and also to PEM his superior during the tenure of the project.

- LE-Process;
- LE-Piping;
- LE-Mechanical;
- LE-Civil;
- LE-Arch;
- LE-Instrument;
- LE-Electrical.

Project Procurement Manager (PPM)

PPM manages the following members:

- Procurement Co-ordinator;
- Inspection and Expediting Engineer;
- Logistics Engineer.

Quality Control Manager (QCM)

During the engineering phase, the function of Quality Manager is limited, and during procurement phase, his functions are also limited to resolving only the major issues in manufacturing/shop fabrication, for which inspection is normally carried out by third-party inspectors like Lloyds, Bureau Veritas, and DNV. Hence, his team is seen as one-man organization. QCM's working location is normally the Home office or a place where engineering and procurement activities are performed.

During the engineering and procurement stage, the tasks predominantly are preparation of the quality assurance plan, audit plan documents, and inspection and test plans, conducting weekly meeting with Owner, and participating along with Project Management Team the weekly progress review meeting with Owner. In relation to Vendors, his functions are management of nonconformity report (NCR) and review of inspection and test plan (ITP) of Vendors. He will also assist the Project Management Team in the appointment of Third-Party Inspection Agencies.

However, when Site construction especially activities like field welding and concreting pick-up, the Site QA/QC Manager takes charge of the activities. Quality audits on the project at periodical levels are arranged by this QCM. At times, when Site quality works are under criticism from Owner, QCM has the responsibility to step in to give advice.

HSE Manager

The function of HSE at headquarters is limited to preparing HSE plans, procedures, philosophies, and undertaking any safety audit of Site at corporate level if demanded by Owner or necessitated by the procedures, while Site-based HSE Manager will be responsible for all HSE at Site. In order to conduct Site audit at corporate level, he is expected to make Site visits from headquarters. HSE Manager has direct channel of correspondence with Site HSE Manager, makes assessment periodically, and reports to CEO at headquarters.

HSE Management monitors and controls the activities in respect of safety, health, and environment.

1.2 PROJECT CONTROL AND ADMINISTRATION

The Project Control and Administration is handled by Project Control Manager (PCM). The key roles and responsibilities are (1) cost control, which includes progress statement and invoice-related tasks; (2) planning and scheduling; (3) document control; and (4) risk management.

1.2.1 Cost Control

Invoicing Procedure is a deliverable, and this is prepared and implemented during the execution phase.

Cost control engineer keeps a detailed breakdown of budgeted cost based on the Contract or as directed by top management who may desire more profit and contingencies than allocated at the time of Contract award. Cash flow curves indicating cash-in and cash-out are maintained monthly basis, and this being confidential is not transmitted through official documentation system but handed over in person to Owner's cost engineer. Balance sheet and profit and loss account if maintained constitutes a state-of-art administration, and this information is not/cannot be shared with Owner.

The cost breakdown of Contract price runs as below. Let us assume 1000 million USD Contract with single currency to design, procure, and construct and commission a Gas Process Plant, which treats natural gas from 20 wells before being sent into gas export of international gas grid pipelines. For the purpose of discussion, let us assume the project is located in a desert far away from towns/cities.

As shown in Table 1.1, the Contract price is broken into two parts: the first is lump sum price and the second provisional sum. Contract price breakdown in detail is shown in Appendix 2.

Normally, Owner introduces provisional sum with an option to increase or decrease the quantity occurring during the execution phase. As the engineering information on the items under provision category is not accurate at the bidding/contracting stage, a significant variation could result in upon the development of detailed engineering, and hence, it is prudent for Owner to have a portion of sum under provisional category in Contract price. This provides flexibility for increase by the application of unit rates provided thereof for the quantity variation and thus eliminates arguments or disputes.

For example, flow lines' line pipe lengths between well cluster and well pad can vary significantly if co-ordinates of the wells are changed during the course of detailed design (flow assurance studies). These items of variation can be easily handled through provisional category price of Contract. For handling provisional category, the Owner keeps additional budget to absorb such additional cost (when increase occurs) without a need to seeking the approval of stakeholders of the company or the board. Lump sum price is basically fixed one, and this cannot be varied

TABLE 1.1
Contract Price Breakdown

Description	USD Million
Lump sum price for the work	950
Provisional price for the work	50
Total[a]	1000

[a] Taxes and duties included.

unless both parties approve any change orders related thereof. If change orders are allowed, the processing time for approval cycle takes as much as 3- to 6-month time. Under the provisional lump sum category, to approve payment for any increase over the Contract specified amount, no amendment of Contract is necessary, while under lump sum, amendment of Contract is required although such increase has been approved through changer order by Owner. Amendment so mentioned when made is to reflect the increased price in revised Contract price. Amendment of Contract is again time-consuming exercise, and this takes at least 3–6 months.

1.2.2 Planning and Progress Control

The deliverables Contactor prepares include:

- 90-day starter schedule for early work activities;
- Work breakdown structure;
- Contract calendar dates, including fixing cut-off dates for progress measuring;
- Progress measurement procedure;
- Detailed Level 3 Network Schedule for Engineering and Procurement;
- Detailed Level 3 Network Schedule for Construction and Pre-commissioning/ Commissioning.

The schedule, especially the Level 3 one, requires loading with direct manpower/ man-hour and quantities, without which it lacks credibility.

Level 1 and 2 Schedule

Level 1 and 2 Schedules are explained in Chapter 26.

90-Day Starter Schedule

As Level 3 Schedule preparation/review/approval cycle requires at least 4–5 months for mega projects, it is recommended to have a simple schedule at the start of the project, and such schedule can be in the form of Gantt chart in excel sheet. This schedule is required to monitor closely those activities which could slip into criticality in the early phase and to take corrective measures. The following are minimum among other activities to be tracked in this schedule.

- Those portions of Site which Owner is unable to handover at the commencement date or zero date;
- Soil investigation (geotechnical);
- Design and construction of temporary facilities (office, canteens, residential camps for staff and workers, laydown area, batching plants etc.);
- Placement of purchase orders for long-lead items;
- Procurement of insurances by parties (Owner, Contractor, Subcontractors);
- Setting up batching plants at Site (for producing concrete);
- Preparation of laydown areas.

Level 3 Schedule

A Level 3 Schedule shows activities of engineering, procurement and construction, and commissioning. Further such schedule shows also Owner milestones, which if not accomplished on time during initial phase are feared to push the project into critical path right from the start. In Level 3, the number of activities is significantly higher than that in Level 2. Level 3 is developed post Contract, while Level 2 is part of Contract Document.

For example, the subsection piping discipline under engineering section may expand to include sub-sub-section activities like issuance of plot plan, general arrangement drawing, isometric, line list, and material specification. Likewise, in construction section, the piping will have sub-sub-section activities such as fabrication, erection, support fabrication, support erection, and field alignment/bolting. In Level 3, the total Plant may be split into several area or units. Under each area/unit, the section, subsection, and sub-sub-section of each discipline will be shown. Apart from above, in Level 3, a general section is added to include items like milestones, interface, third party, quality, HSE, and management and control sections. To efficiently control the schedule and its periodical updates, Level 3 Schedule is split into two parts: the first part is for engineering and procurement, while the second for construction, commissioning, and performance test. The first part should be fixed as early as possible.

Fore convenience, Level 3 is split into two parts: the first one for engineering and procurement, and this is wrapped up within first 5 months of commencement date. The second part is dedicated to construction, pre-commissioning, and commissioning; this is wrapped up when (1) the delivery dates for critical, long-lead, and important items are known from Vendors as committed in purchase order (PO); (2) the layout and General Arrangement Drawing (GAD) are made and the preliminary drawings are available to ascertain the quantities of concrete, steel structure, piping inch-dia welding, electrical and instrument Material Take Off (MTO) etc.

Contractor also gives attention to those activities which Owner is to perform and keeps in place a proper strategy in the preparation of schedule. Let us assume the following three activities are under the responsibility of Owner:

1. Early Access: Early access to Site by Contractor happens without delay soon after commencement date of the project. Any delay hereof delays the start of soil investigation (through bore holes) and consequently the engineering, particularly the design of civil foundations. Delay in early access also could delay the establishment of temporary facilities by Contractor, which in turn could subsequently delay the mobilization of key persons required at the start of project Site. Contractor is expected to designate this task of Owner to be a constraint in critical path.

2. Review Comments: If Owner provides review comments anew in every round of submission of important engineering deliverables, the review delays may push the project silently into criticality without catching anyone's attention and a cure in time. Hence, it is important that activity duration and floats are carefully assigned to create pressure on the

Owner and prevent Owner from indulging in making endless comments in the approval of deliverables, especially the Piping and Instrument Diagrams (PIDs) and layout's first issue.

3. Well-Pads Handover: Well-pads handover here means the completion of drilling and then rigless operation. The Christmas tree contains group of valves, including a wing valve at the tie-in point for the EPC Contractor. Christmas tree is installed after drilling to secure the gas/oil well safe. If drilling work by Drilling Contractor is delayed, then EPC Contractor's progress will be impacted, and this aspect is taken care by the Planning Engineer/Scheduler. After drilling by Drilling Contractor, Owner engages Contractor to perform the rigless operation. Further, it must be noted that normally Owner's department that controls the assets are different from Owner's project team. As such, Owner's project team is placed in difficulties in providing timely real-time status from asset controlling team to Contractor, and this situation most of the times leads to inaccuracies in the periodical update of the schedule by Contractor's Planning Engineer with respect to commissioning activities.

It is known that in a few projects, unfortunately the constructed well pads had become useless as the quantity of gas was found much less than designed or the gas contained CO_2 or mercury content more than that Acid Gas Regeneration Unit (AGRU) or Mercury Removal Unit (MRU) could admit at the inlet. To overcome similar setbacks or mitigate the severity of any adverse impact to the schedule, the best practice advised for the Owner is to complete the drilling and rigless activities of gas wells much before EPC Contractor's detailed engineering commences. Pressure Safety Valve (PSV), often called as choke valve, installed at the wellhead/surface generally has a very limited design margin to absorb variations while at the operation. In other words, these valves were seen incapable of handling those wide variations that rendered gas wells useless in some projects. If new PSVs were engineered to meet the changed gas specification, then a timeline of 12 months would be required for delivery which EPC Contractor cannot accept while he is at Plant start-up.

In view of above, Contractor is expected to designate the tasks of well-pads handover to be a constraint in the critical path.

Level 4 Schedule

Level 4 Schedule is explained in Chapter 26.

1.2.3 Reporting

Contractor prepares the formats of reports, schedules, and look-ahead-schedule reports with the consent of Owner.

In general, the reports that this discipline makes are:

- Weekly report;
- Monthly progress/status report;
- Monthly schedule report.

On how should the content, highlights, and summary section appear, both Contractor and Owner jointly decide, if Contract does not specify any template. See sample in Appendix 3.

1.2.4 Document Control

The documents to be used for the project are developed from CDRL, which stands for Contractor Deliverables Requirement List. Such documents are identified by Contractor at the time of bidding stage in its execution proposal in the technical bid. In some projects, Owner provides a provisional list of such requirements that include philosophies, procedures, plans, and engineering specifications.

Under Level 3, the tasks involved include the preparation of

- Master Document Register;
- Document Control Procedure;
- Document Numbering Procedure;
- Project Procedures List.

Contractor may have his own document control system practiced in his various projects.

AIM software is chosen nowadays for projects of complex nature to effectively manage the data-based document management. In absence of any proper documentation system with Contractor, Owner stipulates the usage of such software for document management.

1.3 QUALITY MANAGEMENT

The tasks under Level 3 Schedule involved are preparation of QA/QC Plan Documents and QA/QC Procedures Documents. At Level 4, the tasks list expands to include several procedures/plans/ITP which Contractor or his Subcontractors prepare. However, Level 4 Schedule preparations are kept under the scope of Subcontractors of the Contractor.

1.4 THIRD PARTY AND AUTHORITIES

1.4.1 Approvals and Certification

Contractor provides a co-ordination procedure addressing roles and responsibilities of Owner, Contractor, and Subcontractor with respect to obtaining various permits and approval from government and/or statutory authorities at various stages of engineering, procurement and construction, and commissioning phases of the project. Co-ordination procedure is a Level 3 activity for Contractor in the schedule.

1.4.2 Permitting and Authorization

This activity/task is related to progress achieved progressively on various permits and approval with update, and this is an ongoing activity until commissioning stage and is reflected in the Level 3 Schedule update.

Considering the work load of tasks, one engineer is assigned, and he is generally referred to as "Authority Engineer". He monitors and controls as part of organization of project management. He prepares a matrix of various approvals with timeline, duly reviewed by Schedule Controller and approved by Project Manager. The Authority Engineer updates monthly the status, which also becomes one of the attachments to monthly status report to be submitted to Owner.

1.4.3 Other Third Parties

This activity/task is an ongoing one from the beginning of the project until completion of equipment and material deliveries and construction at Site. The scope of this task begins first with the engagement of third-party agencies. The appointed third-party agencies inspect equipment and materials at Vendor or fabrication shop. The extent of such inspection is covered in Contract and a dedicated procedure thereof. DNV, Bureau Veritas, Lloyds, and many others alike are known to be providing third-party inspection services.

1.5 INSURANCE SPECIFIC TO WORK

The deliverable under Level 3 is Insurance Management Plan.

The following are insurance applicable to the project. A detailed description is provided under Section II Commercial Aspects.

- EAR (Erection All Risks Insurance);
- Third-Party Liability Insurance;
- Marine Cargo Insurance;
- Vendor Insurances;
- Other insurances.

2 HSE, Social, and Security

2.1 HSE MANAGEMENT AND ORGANIZATION

HSE procedures, HSE plans, and HSE organization tasks are covered under this section of Level 3. HSE management function is described in Chapter 1.

Contractor prepares a HSE organization chart reflecting name, mobile number of safety personnel (of both Contractor and Owner), and photograph.

Contractor's Head office provides support, which is continuous and full-fledged in the early stages, and occasionally later stages.

2.2 HSE AT SITE

The HSE organization is responsible for the management of all activities under HSE and ensures that all workers and staff working in the Site comply with the procedures.

Although HSE plan can be limited to one document in the Level 3 Schedule, HSE procedures can be more in number and this can be captured in Level 4. HSE plans and procedures require approval or review according to Document Approval Procedure, where the document class and approval category is specified. A brief list of deliverables is given below, while Appendix 10 provides a longer list.

- Fire Prevention Procedure,
- Emergency Evacuation Procedures,
- Scaffolding Procedure,
- Night Work and Permit Procedures,
- Confined Space Work Procedures.

2.3 HSE AT OTHER LOCATIONS

The "other location" means work places in Vendor shop or any place away from Site, where some HSE action is required as part of project execution but direct HSE management over the Vendor shops is not intended to be the responsibility of Project Organization. Each Vendor is responsible for any HSE issues arising thereof as per Safety Regulations of the country where the shop is located. Project HSE report does not take into account any loss of man-hours resulting from injury/fatalities or recordable safety incidents occurring in Vendor shops for the monthly calculation of LTIFR and TRIR, respectively.

2.4 SECURITY

Security Co-ordination Plan is a deliverable that addresses the employment of security agencies and deployment of its personnel at entry and exit of the premises including patrol. It further addresses how the Site-specific security ID card is issued and maintained. Security is for men and materials (including installed equipment and instruments, and temporary facility).

Security assumes importance. Where law and order or terrorism is a grave concern, military is deployed at Site and Camps. Normally, on security issues of national importance, Contractor and Subcontractor have limited roles. Some Vendors delay or refuse to dispatch their supervisors to Site where security for its personnel is a matter of concern; such situations sometimes delay commissioning of the Plant.

In a normal situation, the purpose of security is to prevent theft and unauthorized intrusions by outsiders; the Contractor and Subcontractors are responsible for the protection of their properties and belongings with the help of their own security personnel.

3 Services and Facilities to Owner

3.1 AT HOME OFFICE OF CONTRACTOR DURING ENGINEERING AND PROCUREMENT

During engineering and procurement phase, a major contingent of Owner's project team stays in close proximity to the Contractor's in order to deliver an efficient project execution. Close proximity improves significantly the co-ordination among the project teams. With the improved co-ordination, the unity among the teams is easily achieved. As stipulated in Contract, Contractor in his office provides Owner with furnished office, including computers, copy machines, printers, internet and telephones, and stationaries. To the extent possible, often both teams share the same floor of the office. Owner team presence in Contractor office is represented in man-months. If Owner desires to prolong the presence due to project delays without approving any additional man-months for the prolonged stay, then this issue would be a matter of cost concern to Contractor. Since the cost involved is somewhat significant, Contractor gets an opportunity to dissent any prolonged stay without compensation.

For the purpose of providing office to Owner during execution phase, Contractor during the bidding stage specifies the man-months as part of Contract condition related to Owner's team presence in the former's office. These man-months later get reflected in Contract under section called Services and Facilities to Owner, failing which Owner can possibly prolong his stay. Closer and longer Owner team stays with Contractor, higher the efficiency of project execution in engineering and procurement. However, when the question of overruns arises, Contractor does not favor the prolonged stay.

For providing Owner personnel with residential accommodation, these obligations are captured in the Contract during the formation of Contract. If Contract does not include these obligations but Owner requires it post Contract, then an administrative mechanism to have these expenses reimbursed from Owner on monthly basis is put into place. Such administrative mechanism does not fall under change/variations and is known to have been practiced in many projects.

3.2 SERVICES AT SITE

Contractor at Site builds and maintains temporary office, canteen, and residential camps for Owner alongside Contractor's Temporary Facility setup. Internet, walkie talkie radios, transportation, and security facilities are provided.

In countries where security to foreign nationals is an issue of grave concern, then an elaborate military camp and facilities and security systems is contingent.

Camp facilities are accordingly designed and built prior to the mobilization of expatriate engineers/supervisors.

In some projects, due to security reasons, the commutation of Site workers of Contractor and Subcontractors, from their residential area/towns/airports to Site, have been done only with convoys of military providing escort.

4 Engineering

Engineering is subdivided into

- Detailed Engineering (DED);
- Procurement Engineering;
- Construction Engineering;
- Pre-commissioning and Commissioning Engineering;
- Start-up and Performance Test Engineering;
- Operation Engineering;
- Maintenance Engineering;
- Interface Engineering;
- Field Engineering.

Before the start of DED, Contractor verifies the Front End Engineering Design (FEED) Package provided by Owner. FEED contains licensors' basic engineering packages if the project is process-oriented one, and therefore, the possibility for discrepancy between project specifications adopted for EPC phase and the specifications followed by licensors during FEED phase is not ruled out. Furthermore, Owner fears a variation in subsoil strata from the soil data he has gathered during his soil investigation preceding the start of bidding phase. To reject claims that Contractor may bring up on account of above two situations, Owner tends to have the FEED endorsed by Contractor during bidding stage itself. Such tendency is basically a pre-emptive approach. Can the Contractor during the execution phase stand by the endorsement that has been hurried through during the bidding stage? For a proper endorsement to take place, the Contractor should have the expertise (in-house or outsourced) and adequate time during the bidding stage. For fear of disqualification, Contractor comes forward to blindly endorse, and for this blind act and to cover the risk thereof, he loads some contingency cost to the bid price. Addition of contingency reveals the uncertainties that Contractor embraces. Therefore, whatever endorsement takes place during bidding stage fails to live up to the reality. Endorsement pre-empted as such during bidding stage most probably falls under exculpatory clause, which will be discussed in another section. A typical list of deliverables under FEED is shown in Appendix 5.

Contractor prepares a flow chart for DED, as part of a procedure related thereof, to show how review/approval, sequences, and predecessors are addressed. It indicates Owner's participation in the turnover of the key deliverables. A typical deliverable sequence related to material requisition cycle is shown in Appendix 4.

Engineering execution plan specifies a list of software programs to be used during engineering. Normally, such software is part of IT PLAN section of execution proposal in the bidding process, and later, this becomes part of Contract Document.

During the execution, the Engineering Manager understands from the Level 3 Schedule update issued by Project Control department the critical engineering activities where float is zero or reduced to zero. He tracks down the deliverables in critical path. He also tries to find those action delays in respect of review and approval from Owner, which results in increase in severity of criticality. His concern centers on the float that has become zero and furthermore negative. Under no circumstances should both Owner and Contractor top management allow deliverables of critical nature (e.g., Piping and Instrument Diagrams-PID and layouts) suffer rejection when issued for review to Owner. Class 1 documents when rejected by Owner bring to halt those project activities which are related thereof, but when commented doesn't. Halt to critical activity means project delays are allowed to take birth. Deliverables of critical nature are mostly Class 1 documents.

4.1 DETAILED ENGINEERING

DED branches off into the following 14 numbers of sub-DEDs or even more. Lead engineers are assigned for each branch of engineering. According to the organizational practice, one Lead Engineer may handle two or three sub-DEDs. For example, a Lead Engineer process can handle Process Engineering, Flow Assurance and Safety Engineering.

- Process Engineering;
- Flow Assurance;
- Safety Engineering;
- Mechanical Engineering for Rotating Equipment and Packages;
- Material Selection, Corrosion Management, and Surface Protection;
- Mechanical Engineering for Static Equipment;
- Piping Engineering and Layout;
- Instrument and Control System Engineering (ICSS);
- Telecom Engineering, Electrical Engineering;
- Civil Works and Structural Engineering;
- Building and Architectural Engineering;
- Heating Ventilation Air Conditioning (HVAC) Engineering;
- Pipeline Engineering;
- Plant 3D Modeling.

4.1.1 PROCESS ENGINEERING

Process Engineering produces the following deliverables to accomplish its scope under DED. There are 17 deliverables as mentioned here below, and this could vary from project to project, and Owner to Owner. Some of the deliverables mentioned hereunder are mere development from FEED, while others are not.

- Calculation notes (usually issued for compressors, vessels, tanks, basins, control valves);
- Utility balances (calculation note) – this exercise is to understand how utilities such as instrument air, nitrogen, water (for fire, cooling, drinking, etc), steam, hot oil, and electricity are adequately produced from the respective systems and consumed.

- Field life;
- Fuel gas balance (this exercise is required to understand the various grades and quantity of fuel gas required by different equipment, mainly gas turbine generators, drivers for compressors)
- Effluent summary;
- Emergency shutdown logic diagrams;
- Process and utility descriptions;
- Process flow diagrams;
- Heat mass balance;
- General philosophy on process;
- Philosophy on process and utility operation and control;
- Piping and instrument diagrams;
- Process design procedure;
- General process report;
- Pressure safety valve relief loads (calculation note);
- Emergency depressurization (calculation note);
- Utility flow diagrams (UFD).

4.1.2 Flow Assurance Study

Flow Assurance Study Deliverables include Steady-State Flow Assurance Study (basically a calculation note) and Transient State Flow Assurance Study (basically a calculation note); these are related to pipelines design. It may be noted here that piping is different from pipeline, and therefore, codes governing design of piping are different from pipelines. Contract provides a list of applicable codes for carrying out design. For example, ASME B31.3 relates to piping, while ASME B31.4 to pipelines.

Flow Assurance Study for the pipelines (not piping) is developed with help of related field surveys and reports, which are handled through third-party companies, specialized in this field. The Pipeline Subcontractor if engaged by Contractor for EPC execution of pipelines may not be having in-house expertise for performing flow assurance studies. When Owner specifies the names of engineering Subcontractors in the Contract section of Approved Vendors and Subcontractors, Contractor even if he has an in-house expertise has obligation to subcontract such engineering task to the approved Subcontractors.

4.1.3 Safety Engineering

Under this section, as many as 19 deliverables appear. As said earlier, the number of deliverables could vary according to the nature of the project, industry segment, and requirement of Owner.

Safety Engineering Deliverables include the following:

- Hazardous area layout;
- Data sheet (Safety Engineering);
- Data sheet for clean agent system;
- Philosophy (Safety Engineering);
- Report (Safety Engineering);

- Risk assessment report;
- RAM study – reliability, availability, and maintainability;
- Design HSE;
- Process HAZOP and SIL review;
- Performance standard (Safety Engineering);
- Escape and evacuation route plan;
- Fire-fighting equipment layout;
- Fire and gas detection layout;
- Fire-fighting process flow diagram (PFD);
- Fire-fighting piping and instrument diagram (P&ID);
- Fire zone drawing;
- Hydraulic calculation report analysis (fire-fighting);
- Fire and gas cause and effect (Safety Engineering-Fire-Fighting);
- Fire alarm equipment location plan.

4.1.4 Mechanical Engineering for Rotating Equipment and Packages

Rotating equipment and packages deliverables include the following:

- Equipment list;
- Specification mechanical general;
- Vendor print check and review for rotating equipment;
- Mechanical data sheet.

4.1.5 Material Selection, Corrosion Management, and Surface Protection

Deliverables include the following:

- Specification cathodic protection;
- Specification material selection (painting);
- Specification material selection (insulation);
- MSD (material safety data);
- Material and anticorrosion;
- Calculation note (cathodic protection);
- Detail drawing (cathodic protection).

4.1.6 Mechanical Engineering for Static Equipment

Deliverables are mechanical data sheet, Vendor print check, and review.

4.1.7 Piping Engineering and Layout

Deliverables include the following:

- Data sheet (general);
- Plot plan;

- Plot plan review report;
- Specification piping;
- Material specification piping;
- Tie-ins;
- Line list and critical line list;
- Special support drawing;
- General arrangement drawing;
- Isometric.

4.1.8 Instrumentation and Control System Engineering (ICSS)

Deliverables include the following:

- Specification;
- Control room layout (instrumentation);
- Control room wiring plan (instrumentation);
- Main cable plan;
- Main cable wiring plan;
- Connection detail;
- Hook-up diagrams;
- Trouble shooting loop diagram;
- Instrument index;
- Data sheet;
- Instrument cable drum schedule;
- Instrument cable list;
- Instrument connection list;
- Grounding philosophy and diagram;
- ICSS scope of work and services;
- Level sketch;
- Fire alarm system wiring plan;
- Fire alarm system schematic diagram.

4.1.9 Telecom and Electrical

Telecommunication Engineering Deliverables include the following:

- Specification;
- Typical installation detail drawing;
- Philosophy;
- Telecom equipment numbering procedure;
- Telecom wiring plan;
- Telecom room equipment layout;
- Telecom field equipment layout;
- Block diagram;
- Connection and other diagram;
- Cable schedule;
- Telecom equipment list.

Electrical Engineering Deliverables include the following:

- Calculation note;
- Lighting plan;
- Diagram, data sheet;
- Electrical equipment list and load balance;
- I/O list;
- Philosophy;
- Procedure (commissioning and start-up);
- Specification;
- Grounding plan;
- Main cable route plan;
- Electrical equipment layout;
- Power plan;
- Interconnection wiring diagram;
- Single line diagram;
- Cable schedule;
- Hazardous certificate list.

4.1.10 Civil Works and Structural Engineering

Deliverables include the following:

- Drainage plan;
- Manhole schedule;
- Pit and pond plan and detail;
- Specification;
- Soil investigation plan;
- Typical detail drawing;
- Geotechnical investigation report;
- Paving plan;
- Cable trench and duct bank detail;
- Miscellaneous foundations details;
- Material take-off report;
- Drawings for Site preparation;
- Pump foundation design;
- Storage tank foundation design;
- Column foundation design;
- Heat exchanger foundation design;
- Vessel foundation design;
- Calculation sheet (Off-Site);
- Foundation detail (Off-Site);
- Paving plan (Off-Site);
- Bar-bending schedule.

4.1.11 Buildings and Architectural Foundations

Building and Architecture may include equipment structures, pipe rack, miscellaneous structure, shelters, and buildings; for each of which, the following deliverables are produced.

- Calculation sheet;
- Project specification;
- Typical drawing;
- Structural drawing;
- Foundation drawing;
- Bar-bending schedule;
- Material Take Off (MTO);
- Architectural drawing (for building only).

4.1.12 HVAC Engineering

HVAC provides air conditioning to control rooms, laboratories, office buildings, etc in cost-effective way. Deliverables include the following:

- Calculation sheet;
- HVAC drawing;
- P&ID, data sheet;
- Cause-and-effect diagram;
- Specification.

This engineering is mostly subcontracted to the specialized companies as per Owner or Contract requirements. Ducting work is often fabricated at Site and is installed by Subcontractors of Contractor.

4.1.13 Pipelines Deliverables

Pipelines may include flow lines from wells to cluster, trunk lines from cluster to central processing facility, and export lines. The DED of pipelines is subcontracted portion of EPC Subcontract between Contractor and Pipeline Subcontractor. Pipeline being a specialty field, EPC Contractors engage an approved engineering Subcontractor in this field.

- Calculation sheet;
- Data sheet, report;
- Specification;
- Pipeline routing drawing (trunk line);
- Pipeline alignment sheets (trunk line);
- Pipeline layout drawing (flow lines);
- Pipeline alignment sheet (flow lines).

4.1.14 Plant 3D Modeling

Deliverables may include the following:

- 3D CAD implementation plan (piping);
- 30% Model Review;
- 60% Model Review;
- 90% Model Review;
- Close-out report for 30% Model Review;
- Close-out report for 60% Model Review;
- Close-out report for 90% Model Review.

Note: In some projects, Owners include 30% model in FEED stage, while 60% and 90% in DED stage.

4.2 PROCUREMENT ENGINEERING

4.2.1 Central Processing Facility (Main Plant)

Procurement Engineering branches off into various subdisciplines as below for producing deliverables. The issuance of these deliverables falls under various stages of issues such as IFR (issue for review), AFD (approval for design), and IFC (issue for construction).

The deliverables that Contractor prepares include insurance claim procedure, shipping plan, and other procedures. And these cover procurement, shipping, and inspection discipline separately.

Under rotating (general) subdiscipline, the deliverables that Contractor prepares include RFQ (request for quotation) and TBE (technical bid evaluation) for the following:

- Pumps designated under API;
- Pumps not designated under API;
- Centrifugal pumps;
- Positive displacement pumps;
- Vendor package units.

Under static equipment subdiscipline, the deliverables that Contractor prepares include RFQ and TBE for the following:

- Columns;
- Vessels;
- Reactors;
- Heat exchangers (plate heat);
- Heat exchangers (shell and tube);
- Tanks.

Under safety subdiscipline, the deliverables that Contractor prepares include RFQ and TBE deliverables for the following:

- Fire-fighting equipment;
- Safety equipment;
- Clean agent systems.

Under piping subdiscipline, the deliverables that Contractor prepares for underground piping and above-ground piping separately include (1) material take-off, (2) RFQ, and (3) TBE for the following categories of items as are applicable thereof:

- Pipes (CS/SS/alloy: welded or seamless);
- GRE piping (normally, it is for UG piping in some projects depending on soil conditions);
- Forged fitting;
- Wrought fitting;
- Flanges;
- Cast valve (globe/gate/check);
- Dual-plate check valve;
- Ball valve;
- Butterfly valve;
- Foot valve;
- Extended stem valve (ball and butterfly);
- PSV;
- Gaskets;
- Bolt/nut;
- Bug and bird screen;
- Liquid trap;
- Cold support;
- Drip funnel;
- Eye washer;
- Flexible hose/hose connector/quick coupling;
- Sample cooler;
- Silencer;
- Spring support;
- Strainer;
- Flame arrestor.

Under steel structure, the deliverables that Contractor prepares include the material take-off for main pipe rack structures, RFQ, and TBE for equipment structures and steel structure. The material take-off normally happens at three stages of engineering.

Under electrical, the deliverables that Contractor prepares include RFQ and TBE for the following typical items:

- HV switchgear;
- LV switchgear;
- Power distribution panels;
- Power transformers;
- UPS (AC/DC);
- Electrical cables;

- Electrical control system;
- Solar power system;
- Electric heat tracing;
- Cable tray;
- Lighting fixtures;
- Diesel engine generator;
- Cable glands;
- Cathodic protection system;
- Fire alarm.

Under telecommunication, the deliverables that Contractor prepares include RFQ and TBE but the Contractor doesn't have the list of items, and hence, he depends on the telecommunication system Subcontractor. As such, in the RFQ, Contractor merely provides the concept, philosophy, functional requirement, and specifications. Subcontractors in the bidding stage perform pre-bid engineering to prepare quotation (priced and technical). This concept may change from project to project.

Under instrument, the deliverables that Contractor prepares include the 1st Material take-off, 2nd Material take-off, and 3rd RFQ and TBE for the following items:

- Control valves;
- Choke valves;
- ON/OFF valves;
- Self-actuated control valves;
- Breather valve;
- Field instruments;
- Cables;
- Junction boxes;
- Metering system;
- Tank gauging system;
- Analyzer;
- ICSS.

Under HVAC, the deliverables that Contractor prepares include HVAC system RFQ and HVAC system TBE. HVAC system is also usually provided by a Subcontractor/ Vendor who supplies under EPC package until commissioning and performance test. However, such Subcontractor often opts to exclude the installation of field ducting work and related ducting fabrication from its scope.

4.2.2 Off-Site

For deliverables to be produced under General Category, Contractor prepares the RFQ and TBE for the following:

- Isolation joint;
- Coated seamless;
- Coated CS welded pipe;

- Material take-off;
- Barred tee;
- Large radius bends or special bends (to make pipelines piggable).

Under electrical, Contractor prepares the Material take-off at IFI, Material take-off at IFD (issue for design), and Material take-off at IFC stages.

4.3 CONSTRUCTION ENGINEERING

4.3.1 Main Plant and General

The deliverables that Contractor prepares include IFR, IFD, and IFC.

- Calculation;
- Organization chart;
- Manual;
- Plan;
- Procedure;
- Reports.

4.3.2 Off-site

Under Off-Site, the deliverables that Contractor prepares include manual IFA, manual IFD, manual IFC, and manual AFC.

4.4 PRE-COMMISSIONING AND COMMISSIONING ENGINEERING

Under this, the deliverables that Contractor prepares include pre-commissioning dossier (one per system) and commissioning dossier (one per system).

Since the start-up and commissioning of the total Plant takes place much after the completion of commissioning of each subsystem, the Start-up Manager can handle the portfolio of commissioning activities. Normally, pre-commissioning activities are placed under the responsibility of Construction Team, while the commissioning and start-up under the Project Manager. The organization chart attached is self-explanatory.

Table 4.1 provides the split of activities between pre-commissioning and commissioning. However, this split may vary from project to project based on the definitions incorporated in the Contract.

4.5 START-UP AND PERFORMANCE TESTS ENGINEERING

The deliverables that Contractor prepares include start-up plan and performance test plan.

The start-up and performance testing are important activities of the project in its final stage. In oil and gas industry, some Owners may have experienced operators,

TABLE 4.1
Split of Activities Between Pre-Commissioning and Commissioning

Discipline	Activities	Pre-Commissioning	Commissioning
Piping	Hydrotest	X	
Piping	Gross leak test		X
Piping	Flushing and cleaning	X	
Electrical	Solo-run		X
Electrical	Continuity test	X	
Instrument	Continuity test	X	
Instrument	Loop test		X
Instrument	Function test		X
Mechanical	Vessel box up	X	
Mechanical	Final alignment	X	
Mechanical	Hot alignment		X

and hence, they take the driver seat when the gas-in event comes up, although as per Contract, Contractor's work scope includes start-up, stabilization, normal operation mode, and performance test. Performance test is recommended only if the Plant has run continuously without tripping at least 72–96 hours from the start of normal operation, and this condition may vary from project to project.

The organization of start-up team of Contractor syncs with that of the Owner. It is important that Owner show inclination to take control of operations immediately after performance test is conducted upon which the custody transfer of the Plant takes place. Custody transfers from Contractor to Owner.

4.6 OPERATION, MAINTENANCE, AND INSPECTION ENGINEERING

MIEC (Maintenance and Inspection Engineering Contractor) is appointed by Contractor as a Subcontractor to execute all scope of work falling under this section.

The deliverables that Contractor prepares include operating manual, training module-general, training organization and execution plan, draft final training report (50%), and final training report-general.

MIEC's deliverables are:

- MIE execution plan;
- Data collection and compilation;
- Maintenance program and plan/RCM study;
- Inspection programs and plan/risk-based inspection study;
- Integrity engineering, spare parts, and specific tools;
- Computerized maintenance and inspection management systems;
- Maintenance and inspection manuals;
- First-level maintenance, specific maintenance Contract, handover.

4.7 INTERFACE ENGINEERING

Interface Engineering deliverables are not developed in this book. The extent of key deliverables to be generated depends on the nature of the project, especially the existence of tie-ins of the new Plant with old one and others.

4.8 FIELD ENGINEERING

The involvement of Field Engineering can be minimized if a good quality of engineering is done until 60% and 90% Model Review at headquarters of the Contractor. The field piping modifications can be reduced if emphasis is made at headquarters for a thorough check by allowing additional engineering man-hours to engineering group. The piping interface with pipe rack structures and equipment platforms and cable trays routes is properly checked in the model in order to avoid field modifications; this becomes the strategy of Project Manager when the time arrives for issuance of drawings for construction (IFC).

The deliverables are not shown here. It is an activity which surfaces when modification in field is required on the approved drawings. This is administered through a dedicated procedure that lays down rules and formats to be followed.

5 Procurement and Supply

Before addressing the activities under Level 3, the organization, roles, and responsibilities are described.

It is widely known that EPC firms have their core procurement cell, which is meant to function independently and keep confidentially the sensitive information. The procurement cell gives only limited access to Project Management Team. The Project Procurement Manager, PPM, is the interface between this cell and Project Management Team of Contractor. PPM will have under him a coordinator, team leader for procurement and expediting, team leader for inspection, and a team leader for logistics. PPM attends all Project Weekly Meeting and Monthly Meeting in addition to his attending Procurement Meeting weekly with Owner's counterpart.

Contractor procures and supplies the equipment and materials from approved Vendors and suppliers. Normally, Approved Vendor List is followed. However, Contractor will face uphill task if he is not given the flexibility to expand this list during execution stage. This expansion of list may assume significant importance due to reasons that

1. Some approved Vendors may be running with full load at the time of Contractor's seeking quotes and participation;
2. Contractor may have the possibility to optimize cost from equally good Vendors who may not appear in the approved Vendor list;
3. Some approved Vendors may be at the verge of bankruptcy due to market situation or internal issues;
4. Contractor may have bad experience with a few in the approved list.

In view of above-said circumstances, Contractor can have a strategy and seek expansion of Vendor list. Such expansion is possible if he persuades Owner earnestly by providing him with prequalification dossiers accompanied with audited financial statements (balance sheet, P&L statement at least for the past 3 consecutive years). Prequalification document should also specify how this Vendor has successfully supplied a similar equipment to various clients of repute in the past. Excellence in quality assurance and control in the Vendor's organization can be an added advantage in this attempt.

Procurement strategy for buying mechanical seals assumes importance when the maintenance aspect during operation is considered. In case of procurement of mechanical seals, which is part of pumps and compressors, Contractor can preselect a Vendor for supplying mechanical seals beforehand and enter into a frame agreement so that his product (mechanical seal) may be used in pumps and compressors that Contractor later buys from other Vendors. Such strategy gives the following advantages:

- Reduction in initial cost (capex);
- Better quality;

- Maintenance is made easy by entering into a frame agreement with a single supplier, and thereby removes administrative hassles in handling multiple suppliers during shutdown maintenance;
- Reduction in operation and maintenance cost.

Procurement strategy for negotiation also assumes importance. If Vendor understands that Contractor is in a flux of urgency, the former may not provide the discounts which may come in reasonable values. The procurement of LLI and critical equipment should have strategy to have more time for negotiation. If the purchase order cannot be placed for such equipment as per deadline stated in Level 3, Contractor should not hurry up and place order for value that may exceed the budgeted cost significantly. No Owner will compensate the Contractor for those additional costs, which exceeds budgeted and proposal basis. In any project, there will be some possibility for time extension as concurrent delays are inevitable, and for such concurrent delays, the reasons could be traced to both Owner and Contractor. Hence, all these factors in a whole picture should be taken into account if Contractor is under stress due to higher price in the market for the product in comparison with his proposal basis.

Material requisition is a technical document which contains all the technical requirements of the equipment, materials, or services, including the inspection and test requirements, and documents and key deliverables to be issued by Vendor. Material requisition undergoes revisions as an when any technical changes take place during the execution. Material requisition is, inter alia, a part of purchase order issued by Contractor to Vendor. The procurement team closely monitors the revision of material requisition and evaluates the impact on the ongoing manufacturing in the Vendor shop.

In general, five common activities dominate in Level 3, which are explained below, related to procurement of items grouped under each material requisition.

Five Common Activities under Level 3

i. Receiving quotation from Vendors/suppliers;*
ii. Performing commercial bid evaluation of the received quotations, CBE;
iii. Issuing purchase order to the successful Vendor/supplier, PO;
iv. Receiving the Vendor print (key deliverables to be submitted to Owner and for internal use);
v. Fabrication or manufacturing that takes place under the Vendor premises/ shop.

* For obtaining quotation, the deliverable RFQ is used, which consists of material requisition, general terms and conditions wherever relevant, particular terms and conditions. The general terms and conditions adopted for Vendor align with the terms and conditions in the Contract between Contractor and Owner. The quotation is invited in two parts: technical and commercial. Commercial quote is received in sealed envelope and is opened among the team members relevant thereof. Where sealed cover is not practicable in the era of electronic mails, password-protected files are used. Upon qualifying the technical part through TBE and making alignment of TBEs across all bidders, priced bids envelopes are opened. Technical bid also contains unpriced quote. Unpriced section gives an idea to Contractor how bidders have formed the quoted price. Vendor is advised to strictly follow the format or table for quoting the price reflecting how various taxes and duties that bear on the quoted price; otherwise, it will be difficult for Contractor to compare the quotes across the bidders.

Procurement activity can be broken down into the following subactivities:

- CI 5.1: Supplier Assistance during Construction/Commissioning/Start-up;
- CI 5.2: Supply of Critical and Long Lead Items;
- CI 5.3: Supply of Equipment and Materials (excluding 5.2 above) for Main Plant;
- CI 5.4: Supply of Equipment and Materials for Off-Site;
- CI 5.5: Consumables and First Fills.

5.1 SUPPLIER ASSISTANCE

Contractor plans the deployment schedule of supervisors/specialists in such a way as to minimize the unnecessary standby cost and delay in obtaining visas.

The standby unit rates (per hour or per day) can be as high as 150–200 USD/hour depending on the supplier and its location.

In some instances, Vendor supervisors fail to arrive on time when the Plant is ready for commissioning and performance test runs. Owners sometimes do not accept performance test of any package or equipment if the respective Vendor supervisor is absent during the tests. Such situations could in turn result in delays to the plant performance test, which if not achieved on time renders the Contractor liable to pay liquidated damages.

Since the Vendor supervisor listed for construction phase cannot be competent for subsequent pre-commissioning and commissioning phase, a detailed plan of mobilization to address this issue is made at Level 3 or Level 4 Schedule. The per diem cost for Vendor supervision is usually substantial. Longer the stay in Site, greater is the revenue to Vendors but for Contractor it is a matter of cost overrun and concern. Further, Contractor needs Vendor supervisors to train the operating personnel of Owner in two phases. In view of above, Contractor gives due attention to have Vendor supervision schedule split as below along the timeline.

- Assistance to installation and construction;
- Assistance to pre-commissioning and commissioning;
- Assistance to start-up and performance tests.

5.2 SUPPLY OF CRITICAL ITEMS

Under critical items, the following five common activities appear per equipment listed under engineering section.

Common Activities/Tasks

i. Receiving quotation from Vendors/suppliers#;
ii. Performing commercial bid evaluation of the received quotations;
iii. Issuing purchase order to the successful Vendor/supplier;
iv. Receiving the Vendor print (key deliverables to be submitted to Owner and for internal use);
v. Fabrication or manufacturing that takes place under the Vendor premises/shop.

Critical Items/Tasks

- Integrated Control and Safety Systems (ICCS), which include data management and condition monitoring, remote terminal units to support Off-Site telemetry under Instrument discipline;
- Telecom system Contractor supplied items;
- Long lead items, LLI (MP compressor package) under Mechanical Package discipline;
- LLI (Turbo Expander Package, etc) under Mechanical Package discipline.

LLI list varies from project to project. Items that require long duration for manufacturing fall under LLI. Other critical equipment includes CO_2 absorber, amine regenerator, and GTG.

In a refinery project, this kind of LLI and critical items could include cladded columns, boilers, reactors, and reformer convection modules.

Table 5.1 shows how the five common procurement activities appear per equipment in the Level 3 Schedule, when two equipment are selected to cite an example. The rows required in the schedule are five. If the critical equipment are 7, then 35 rows will appear under Supply of Critical Items.

5.3 SUPPLY OF EQUIPMENT AND MATERIALS MAIN PLANT

5.3.1 Main Plant Equipment

The deliverables or activities are again split across disciplines. This section does not include bulk materials to be procured under each discipline. The disciplines are as follows:

1. Rotating equipment;
2. Stationary equipment;
3. Instrument;
4. Electrical;
5. HVAC;
6. Fire-fighting.

The list of items appearing under each discipline varies as much as the process and utility block diagrams do, and hence, this should be taken as sample case only.

1. ROTATING

   Under Rotating equipment discipline in Level 3 Schedule, the following five common activities/tasks appear per equipment listed under engineering section.

   Common Activities/Tasks

   i. Receiving quotation from Vendors/suppliers#;

TABLE 5.1
Five Common Activities for Compressor/Turbo-compressor equipment procurement

Activity ID	Activity Name	Original Duration Days	Start Date	Finish Date
XXXXXXX01	Receiving quote (MP compressor Pkg)			
XXXXXXX02	CBE (MP compressor Pkg)			
XXXXXXX03	PO (MP compressor Pkg)			
XXXXXXX04	Key VP receiving (MP compressor Pkg)			
XXXXXXX05	Fabrication (MP compressor Pkg)			
YYYYYYY01	Receiving quote (turbo-compressor Pkg)			
YYYYYYY02	CBE (turbo-compressor pkg)			
YYYYYYY03	PO (turbo-compressor Pkg)			
YYYYYYY04	Key VP receiving (turbo-compressor Pkg)			
YYYYYYY05	Fabrication (turbo-compressor Pkg)			

ii. Performing commercial bid evaluation of the received quotations;
iii. Issuing purchase order to the successful Vendor/supplier;
iv. Receiving the Vendor print (key deliverables to be submitted to Owner and for internal use);
v. Fabrication or manufacturing that takes place under the Vendor premises/shop.

Rotating Equipment

Equipment as listed in engineering section is repeated here below. Package equipment is also included under this group.

- API centrifugal pump (BB – between bearing);
- API centrifugal pump (OH – overhung);
- API centrifugal pumps (VS – vertically suspended);
- Non-API centrifugal pumps (VS);
- Positive displacement pumps (reciprocating);
- Positive displacement (rotary);
- Fire water pumps;
- Down-hole pumps;
- Instrument air and dryer package;
- Condensate loading system;
- Acid gas regeneration unit filtration package;
- Tri-ethylene glycol regeneration package;
- Produced water treatment package;
- Chemical injection package;
- Nitrogen generation package;
- Diesel fuel filling station;

- Crane and hoists;
- Mercury removal unit;
- Sewage package under Main Plant;
- Reverse osmosis package.

API pumps are more expensive than ANSI pumps. Aggressive applications determine the requirement of API pumps. Contractor cannot decide API or non-API, as such requirements are generally coming from Front End Engineering and bidding-stage conditions, specifically known as "design basis".

2. STATIONARY

Common Activities/Tasks

Under Stationary equipment discipline in Level 3 Schedule, the following five common activities appear per equipment listed below.

Common Activities/Tasks

i. Receiving quotation from Vendors/suppliers#;
ii. Performing commercial bid evaluation of the received quotations;
iii. Issuing purchase order to the successful Vendor/supplier;
iv. Receiving the Vendor print (key deliverables to be submitted to Owner and for internal use);
v. Fabrication or manufacturing that takes place under the Vendor premises/shop.

Equipment (as listed in engineering section is repeated here below)

- Waste heat recovery unit;
- Flare stack and vent package;
- Thermal oxidizer;
- Columns;
- Column internal;
- Vessel type I;
- Vessel type II;
- Reactors;
- Heat exchangers (plate heat);
- Heat exchangers (shell and tube);
- Heat exchanger (air-cooled);
- Electric heaters;
- Tanks (storage);
- Pig launcher and receiver;
- Filters.

3. ELECTRICAL

Under Electrical discipline in Level 3 Schedule, the following five common activities appear per equipment listed below.

Common Activities/Tasks

i. Receiving quotation from Vendors/suppliers#;
ii. Performing commercial bid evaluation of the received quotations;
iii. Issuing purchase order to the successful Vendor/supplier;
iv. Receiving the Vendor print (key deliverables to be submitted to Owner and for internal use);
v. Fabrication or manufacturing that takes place under the Vendor premises/shop.

Items listed in engineering section are repeated here below, each of which has the five common activities or tasks in the Level 3 Schedule.

- HV switchgear;
- LV switchgear;
- Power transformers;
- Electrical control system;
- AC/DC UPS;
- Diesel engine generator;
- Cathodic protection system;
- Fire alarm.

4. INSTRUMENT

Under Instrument discipline in Level 3 Schedule, the following five common activities appear per equipment listed under engineering section:

i. Receiving quotation from Vendors/suppliers;
ii. Performing commercial bid evaluation of the received quotations;
iii. Issuing purchase order to the successful Vendor/supplier;
iv. Receiving the Vendor print (key deliverables to be submitted to Owner and for internal use);
v. Fabrication or manufacturing that takes place under the Vendor premises/shop.

The five common activities apply to the following items in Level 3 Schedule:

- Metering system;
- Tank gauging system;
- Analyzer.

5. HVAC

HVAC system is another category of equipment for which five common activities as said above apply.

6. FIRE-FIGHTING

Under fire-fighting discipline in Level 3 Schedule, the following five common activities appear per equipment listed under engineering section:

i. Receiving quotation from Vendors/suppliers;
ii. Performing commercial bid evaluation of the received quotations;
iii. Issuing purchase order to the successful Vendor/supplier;
iv. Receiving the Vendor print (key deliverables to be submitted to Owner and for internal use);
v. Fabrication or manufacturing that takes place under the Vendor premises/shop.

For each of the following equipment, the above-said five common activities appear in Level 3 Schedule.

- Fire-fighting equipment;
- Safety equipment;
- Fire-fighting mobile equipment;
- Clean agent system.

5.3.2 Main Plant Bulk Materials

The deliverables or activities are again split across disciplines. Bulk materials are applicable only to the following four disciplines:

1. Piping;
2. Steel Structure;
3. Electrical;
4. Instrument.

1. PIPING

Under Piping discipline in Level 3 Schedule, the five common activities appear per each bulk items listed below for tracking supplies:

- Pipes (CS/SS/alloy: welded or seamless);
- GRE piping (normally, it is for UG piping in some projects depending on soil conditions);
- Forged fitting;
- Wrought fitting;
- Flanges;

- Cast valve (globe/gate/check);
- Dual-plate check valve;
- Ball valve;
- Butterfly valve;
- Foot valve;
- Extended stem valve (ball and butterfly);
- PSV;
- Gaskets;
- Bolt/nut;
- Bug and bird screen;
- Liquid trap;
- Cold support;
- Drip funnel;
- Eye washer;
- Flexible hose/hose connector/quick coupling;
- Sample cooler;
- Silencer;
- Spring support;
- Strainer;
- Flame arrestor.

2. STEEL STRUCTURE

Steel Structure can apply to pipe rack, and individual equipment supporting structures. If the whole of the structures is given to a Vendor/Fabricator for a scope of work covering detail design and fabrication and supply, then such item is to be denoted as one task as Steel Structure in Level 3 Schedule. The five common activities as said above repeat for each PO issued under this Steel Structure discipline.

3. ELECTRICAL

The five common activities repeat for each of the bulk items in the Level 3 Schedule:

- Cables;
- Cable trays;
- Cable glands;
- Lighting fixtures;
- Heat tracings.

4. INSTRUMENT

The five standard activities repeat for each of the following bulk items in Level 3 Schedule:

- Cable glands;
- Cable;

- Cable tray;
- Junction boxes;
- MCT block (multi-cable transit block);
- Tubes;
- Tube fittings;
- Flame and gas detector;
- Ultrasonic flow meter;
- Vortex flow meter;
- Level gauge (magnetic);
- Pressure gauges;
- Temperature gauges;
- Transmitter (radar-level type);
- Pitot tube-type flow elements;
- Major instruments;
- Orifice plates and flanges;
- Valves (on/off);
- Valves (breather);
- Control valves (globe);
- Control valves (self-actuate);
- Rotometer;
- Temperature assembly (Rtd/Tc).

5.4 SUPPLY OF MATERIALS AND EQUIPMENT OFF-SITE

Off-Site Equipment: WHCP/HPU and solar power system. Solar power system is essential for locations which are remote and where power transmission lines are not available. The five common activities repeat for each item.

Off-Site Bulk Materials: Barred tee, bend, coated carbon steel seamless pipe, coated welded carbon steel pipe, isolation joints, and valves. The five common activities repeat for each item.

5.5 CATALYSTS, CONSUMABLES, FIRST FILLS

Some chemicals/consumables/fills are required from the start-up until PAC certificate is achieved. Following are the typical items that Contractor procures and supplies from Vendors or licensors approved by Owner. The five common activities repeat for each item in the Level 3 Schedule.

- Scrubbing agent for acid gas removal system/unit;
- Catalyst for mercury removal system/unit;
- Hot oil for hot oil system/unit.

6 Transport, Logistic, and Storage

6.1 LOGISTIC MANAGEMENT

A deliverable that Contractor keeps in place is Logistic Management Plan in Level 3. It addresses in detail the following important items:

- Heavy and oversized cargoes;
- Sensitive materials/instruments/equipment on which government has imposed restrictions.

6.2 TRANSPORTATION

As per the conditions assumed with respect to the scope of work between Contractor and Vendor, the transportation (task of Contractor) is meant to cover the voyage from the point of FOB until the Site. Vendor transports from his Ex-works until FOB by himself. The cargoes in such transportation segment can cross water and roads. In some projects, Vendor's scope stops at FCA instead of FOB, in which case the Vendor is required to deliver the cargoes (consignment) cleared for export but store it in his premises or at location specified by Contractor before the ocean transport takes place. INCOTERMS provides definitions and battery limit, which is universally followed. Transportation period includes the time required for customs' clearance, which may vary from 30 to 60days. Clearance time is less at ports that have better infrastructure and hassle-free procedures.

The overall duration considered worldwide in projects for transportation is 90days. Ninety days is from FOB to Site.

The deliverables are split across the following nine disciplines:

1. Stationary

   The list of items that should appear in Level 3 Schedule will be the same as they are appearing under procurement section, and hence, this list is not repeated here.

2. Rotating

   The list of items that should appear in Level 3 Schedule will be the same as they are appearing under procurement section, and hence, this list is not repeated here.

3. Electrical

   The list of items that should appear in Level 3 Schedule will be the same as they are appearing under procurement section, and hence, this list is not repeated here.

4. Telecommunication

   Telecommunication items/gadgets, although many in numbers, are considered here as one package item procured from telecom Subcontractor/Vendor, who is responsible for designing and procuring and supplying all items required for the project.

5. Instrument

   The list of items that should appear in Level 3 Schedule will be the same as they are appearing under procurement section, and hence, this list is not repeated here.

6. Structural Steel

   The list of items that should appear in Level 3 Schedule will be the same as they are appearing under procurement section, and hence, this list is not repeated here.

7. Piping

   The list of items that should appear in Level 3 Schedule will be the same as they are appearing under procurement section, and hence, this list is not repeated here.

8. HVAC

   The list of items that should appear in Level 3 Schedule will be the same as they are appearing under procurement section, and hence, this list is not repeated here.

9. Fire-Fighting

   The list of items that should appear in Level 3 Schedule will be the same as they are appearing under procurement section, and hence, this list is not repeated here.

10. Pipeline

    The list of items that should appear in Level 3 Schedule will be the same as they are appearing under procurement section, and hence, this list is not repeated here.

6.3 RECEIPT AND STORAGE TRANSPORTATION AND IDENTIFICATION

Under Level 3 Schedule, this activity is shown as one single activity, although it encompasses all those equipment and materials that are supplied under procurement section. For this one single activity, the start date is effective date of Contract or zero date, while the finish date is the date by which all the erection/installations are completed, as denoted by a key date in the schedule. When materials are received at Site, unpacking and inspection is done to assess the damage. Storage procedure and field material management procedure are part of procedures under Construction, not under Procurement, and this is shown in Appendix 10.

7 Construction

7.1 QUALIFICATIONS

Qualifications mean performance of various tests at Site prior to carrying out work at Site. There are four types of tests well known as follows:

1. Concrete Test: In case of concreting before the first pour on foundation, it is necessary to evaluate through laboratory the concrete strengths of test productions from batching plants meant for the project. It is recommended to conduct such tests at least 3 months prior to using first concrete in construction.
2. Welding Test: It is mainly of piping and pipelines: procedure qualifications (PQR), welding procedure specifications (WPS), and welder qualifications (WPQT).
3. UT/DP/PMI/etc.: Steel plates and sections prior to being used in fabrication undergo magnetic particle test, dye penetrant test (DP) or ultrasonic tests (UT) to detect defects in manufacturing or damages caused in transportation/handling.
4. To Perform Positive Material Identification (PMI), a portable instrument is used. By mere touch of the instrument on the metal to be identified, the chemical composition of the metal gets displayed, and this helps Contractor identify the alloy steel piping materials for storing them separate from other steel materials.
5. Calibration Test: Instruments and equipment used for construction purposes are calibrated at Site, and this calibration is different from the calibration of instruments to be installed in the Plant permanently.

Batching plant cannot be allowed to produce the first test production concrete unless calibrated and certified by local authorities.

Each discipline prepares his qualifications under the guidance and advice of the QA/QC team at Site.

It may be noted here that under Quality Management, the Level 3 activities include procedures and plans, and these activities are to be rolled up within 2–3 months from commencement date, while the activities stated under Construction Level 3 stand different from what comes under Quality Management Level 3.

For the purpose of Level 3 Schedule, all these tasks are bundled into one activity or task named welding qualifications and Non-destructive test (NDT). Qualifications task/activity continues to appear in the Level 3 Schedule until field welding of piping is completed.

In a project of oil and gas, refinery or petrochemical, welding qualifications and NDT are prominent and assume significance in the activities of QA/QC discipline. A detailed working plan in Level 4 or 5 platforms from the respective discipline is required to address the qualifications.

7.2 MAIN PLANT CONSTRUCTION

7.2.1 Preparatory/Early Works

The split of subcontracting for construction work follows what is known as subcontracting plan. Issuance of RFQ (request for quotation)/ITB (invitation to bid), receipt of quotation, and bid evaluation are activities that precede the Subcontract agreement between Contractor and Subcontractor. For executing these tasks, ITB document preparation begins as early as possible, but no later than 2–3 months following the commencement date of the Contract.

To issue ITB/RFQ, an important document known as GTC (General Terms and Conditions) is required, which being a massive document and requiring comments from legal and contract department takes 2–3 months for preparation and hence the write-up work on this GTC document commences along with commencement date of Contract. GTC document becomes part of any purchase order or Subcontract which Contractor may award to Vendor or Subcontractor.

The subcontracting tasks under early work are as follows:

- Subcontracting for Site preparation;
- Subcontracting for soil investigation;
- Subcontracting for work of water well (water well is required if the project is located in an area which is desert or lacks water bodies).

The Subcontracting tasks under main construction work are as follows:

- Subcontracting of civil building work;
- Subcontracting for civil and architecture work;
- Subcontracting of mechanical/piping/steel structure work;
- Subcontracting of electrical and instrument work;
- Subcontracting of painting and insulation.

Other miscellaneous tasks that fall under Early Works execution are as follows:

- Site access, to be obtained from Owner;
- Water well construction;
- Investigation of Site conditions;
- Identification of barrow pits;
- Permit for borrow pits from authorities;

- Mobilization for Main Plant preparation;
- Site preparation;
- Soil investigation.

To perform soil investigation, geotechnical surveys are undertaken by carrying out bore holes and making the report. Although Owner does soil investigation at FEED stage and provides this report as part of SOIL DATA in the ITB, Contractor after commencement date carries out additional soil investigation to verify the SOIL DATA that Owner furnished in the bidding stage. Contractor performs such activities as part of Early Works which appear in the "90-day STARTER SCHEDULE" that precedes a Detailed Level 3 Schedule. Soil investigation consists of two distinct activities: first is bore holes, while second analysis of soil taken out through boreholes and preparing a report. In general, agencies who undertake analysis do not undertake boreholes; hence, two different Subcontractors are engaged by Contractor. Boreholes are Site-oriented work, while the analysis Off-Site work.

Underground mapping is also carried out as part of early work. Early works also include design and construction of camps (for staff and workers) and office for Contractor and Owner. Subcontractors are engaged for performance of these works.

In some projects, the work may involve additionally the brown field activities. The brown field activity means any activity related to modifying or revamping existing units. Green field activity refers to a new Plant to be built, which does not involve any work on the existing Plant. When brown field activities are included in the work scope of Contractor, then the Early Work also includes the measurement thickness of existing piping and health check studies through laser scanning and NDT. If piping wall thickness reduces over the period of operation of the existing units, then the pressure rating is to be reviewed and possibly reduced. If pressure rating is reduced, then its impact on the new Plant design should be understood before proceeding with the detailed design related to Green Field Project.

7.2.2 Foundations (Civil Works Above-Ground and Underground)

The Main Plant is subdivided into seven areas for the ease of execution and monitoring the work. The area is demarcated on the overall plot plan and the respective unit plan. A process block diagram given in Appendix 6 sheds some light on how the process part of the Plant is divided:

1. Reception Area;
2. AGRU and Process Area;
3. Gas Export and Storage Area;
4. Utilities Facilities (which may also spread across Process Areas);
5. Building Area;
6. Flare Zone;
7. Pipe-Rack Area.

(Note: Grouping is typical for gas processing Plant.)

1. Reception Area Civil Work

The reader should note that the list of foundations and the tasks related thereof under this area is not shown but should expect that such foundations will be listed in the Level 3 by the Scheduler or Planning Engineer assigned to the project. However, typical civil works and the tasks that fall under this area are reflected as following:

- Foundations of various equipment that fall under this area;
- Foundations of platforms which may be related to equipment or not;
- Valve box;
- Pits;
- Manholes;
- Trench;
- Duct bank;
- Civil work related to electrical works;
- Civil work related to instrument works.

2. AGRU and Process Area Civil Work

- Same as what is stated under Reception Area above

3. Gas Export and Storage Area Civil Work

- Same as what is stated under Reception Area above

4. Utilities Facilities Area Civil Work

- Same as what is stated under Reception Area above

5. Building Area Civil Work

- Same as what is stated under Reception Area above. In the Building Area, activities shown exclude the buildings, since the buildings are covered under structure. In Building Area, only some portions are building, and the rest is surroundings in which duct bank, trench, manholes, etc. exist.

6. Flare Zone Civil Work

- Same as what is stated under Reception Area above.

7. Pipe-Rack Area Civil Work

- Same as what is stated under Reception Area above.

7.2.3 Structure Installation

This section reflects various structures and buildings.

In this project, no modularization concept is adopted for the fabrication of steel structures.

In pipe racks, several air coolers are installed atop the structure. Piping, instrumentation, and electrical works are carried out simultaneously on the pipe rack at various levels. Due to simultaneous work activities, congestion takes place in working environment, which in turn gives rise to safety issues. For planning modular-type construction, which if adopted can ease congestion and safety issues at Site, both Owner and Contractor are required to agree before the detailed engineering work commences. The elevated location of coolers and heat exchangers has hydraulic relationship with the elevation/proximity of nozzles at which these piping networks enter/exit pressure vessel equipment.

1. Reception Area

    The list of structures under this area is not provided. In any project, the Scheduler/Planning Engineer compiles this list to make the schedule in Level 3. His task becomes easy if he populates the list based on the following categories with the help of respective discipline engineers and equipment list. If three different types of equipment fall under this area, then there will be three tasks under platform, namely, platform erection of equipment 1, platform erection of equipment 2, and platform erection of equipment 3. The names of such equipment are not specified as this is not an important aspect for this book.

    - Pipe supports erection,
    - Platforms erection,
    - Local supports installation,
    - Fire proofing work,
    - Sunshades installation.

2. AGRU Area

    - Same as stated in Reception Area above

3. Gas Export and Storage Area (GESA)

    - Same as stated in Reception Area above

4. Utilities Facilities (Which May Also Spread across Process Areas)

    - Same as stated in Reception Area above

5. Buildings Area

    If any equipment stands in the Building Area, then the description given in Reception Area applies. Additionally, the building activities are generally related to whatever buildings stand in this area. Typically, the following buildings are required:

- Control room;
- Substation;
- Laboratory;
- Technical building;
- Office buildings;
- Fire station;
- Guard houses;
- Workshop;
- Warehouses.

The following activities are typical. Such activities apply to each and every building falling under Building Area, and the list of buildings could vary from project to project. If, for example, three buildings are covered in this area, then for each building the following typical activities should appear in the Level 3 Schedule.

- Excavation;
- Foundation;
- Concrete wall and roof;
- Block work (brick work);
- Plastering work;
- Door and windows;
- Finishing work for walls;
- Finishing work for ceilings;
- Finishing work for floor;
- Structural erection;
- Cladding/insulation for copper tubing if HVAC equipment is present;
- Structural fence work around the HVAC equipment (Note 1).

Note 1: HVAC equipment are located basically outside of the technical building but very adjacent, and hence, HVAC section becomes part of technical building. In other projects, it can be housed possibly inside the building either in cellar or at one side.

6. Flare Zone

- Same as stated in Reception Area above

7. Pipe-Rack Area

- Same as stated in Reception Area above

7.2.4 Installation of Equipment

Some equipment can be oversized and heavier. To handle the logistics and installation of these equipment, there is a need to have a separate Subcontract with heavy rigging companies who own higher-capacity cranes. At the formation of Level 3 Schedule itself, such heavy rigging concept is captured. In order to perform heavy rigging, a heavy duty bearing temporary roads is made. Any pipe rack or structures that may obstruct the movement of heavier cranes are normally kept in abeyance until heavy rigging is completed. Level 3 Schedule indicates such activities whose erection is deferred to later dates.

Equipment are classified into five or more categories as follows:

- Critical and long lead equipment (LLI);
- Equipment which are supplied as part of Vendor Packages but without license requirements (say chemical dozing units – methanol injection);
- Equipment which are supplied as part of Vendor Packages with license involved (e.g., PSA – pressure swing adsorber unit);
- Equipment which are general types (pressure vessels, pumps, exchangers like air coolers or shell and tubes, tanks fully fabricated at shop and supplied to Site without any requirement of field assembly and welding);
- Tanks which are field erected in parts, for which Site-based fabrication and assembly are required further to some prefabrications done at Vendor shops.

As said earlier in this book, the Level 3 activities are monitored area-by-area during construction. The whole Plant is split into the following seven areas to monitor the construction activities.

1. Reception Area

The Scheduler lists out here various equipment that fall under this area. The Level 3 task is to reflect installation; in other words, erection (which includes alignment and grouting). Let us assume equipment are named as X1–X10, then the number of tasks expected is as follows. The categories of equipment shown are examples, and hence, it may vary from project to project.

- Pump X1 installation;
- Drum X2 installation;
- Vessel X3 installation;
- Column X4 installation;
- Internals for Column X4 installation: internals shall be treated as a separate activity;
- Tank X5 installation;
- Filters X6 installation;
- Air coolers X7 installation;
- Condensers X8 installation;
- Package item X9 installation;
- Compressors X10 installation.

2. AGRU and Process Area

- Same as stated under Reception Area

3. Gas Export and Storage Area

- Same as what is stated under Reception Area above

4. Utility Facilities

- Same as what is stated under Reception Area above

5. Building Area (central processing zone)

- Same as what is stated under Reception Area above

6. Flare Zone

Since Flare Zone is a restricted area and is affected by heat radiations, equipment are not located in this area except the following items that are basically part of flaring equipment. The heat radiations spread from flaring. The flaring takes place at the flare tip mounted on the flare stack. The intensity of flaring is high when a sudden release of process gas takes place following operational upsets. Operational upsets are more frequent during Plant start-up and commissioning than normal operation.

- Burn pit installation;
- HP flare tip installation;
- LP flare tip installation.

7. Pipe-Rack Area

- Same as what is stated under Reception Area

7.2.5 Piping/Pipelines

1. Reception Area

- U/G piping work;
- Shop fabrication (AG piping CS);
- Shop fabrication (AG piping SS);
- Shop fabrication (pipe support);
- Field erection (AG piping large bore CS);
- Field erection (AG piping large bore SS);
- Field erection (AG piping small bore CS);
- Field erection (AG piping small bore SS);
- Field installation (pipe supports);
- AG thread piping;
- AG piping bolting.

(Note: CS (carbon steel), SS (stainless steel), AG (above-ground), UG (underground).)

2. AGRU and Process Area
    - Same as what is stated under Reception Area above
3. Gas Export and Storage Area
    - Same as what is stated under Reception Area above
4. Utilities Facilities (which may also spread across Process Areas)
    - Same as what is stated under Reception Area above
5. Building Area
    - Same as what is stated under Reception Area above
6. Flare Zone
    - Same as what is stated under Reception Area above
7. Pipe Rack Area
    - Same as what is stated under Reception Area above

7.2.6 Roads

The activities such as (a) fence and gate, (b) gravel and concrete paving, and (c) road work (un-asphalted and asphalted) repeat in Level 3 Schedule for seven areas as aforesaid.

Road work items are the last items to finish in the project. Road work takes longer duration to complete, and hence, Owner should expect this to finish at the time of Plant Provisional Acceptance. Owner, in a strict contractual sense, has the right to have all items, including roads, pavement, landscaping completed prior to issuing Plant Provisional Acceptance Certificate. However, in practice, this is an impossible condition for Contractor. Hence, both parties should consider this under exception to go ahead with Plant Performance Test Runs and issuance of Plant Provisional Certificate provided such items if not completed are not supposed to create obstruction to start-up and completion of test runs and normal operation. Through a punch management procedure, such exceptions can be assigned a category. Punch items normally have four categories: A, B, C, and D.

7.2.7 Field Instrumentation Cabling

The activities include the following:

1. Cable termination;
2. Secondary cable laying;
3. Junction box;
4. Field box;

5. Analyzer installation;
6. Instrument installation;
7. Fire and gas detection installation;
8. Air pipe;
9. Tubing.

And all these activities are shown again under seven areas as aforesaid in the Level 3 Schedule.

7.2.8 Electrical and Cabling

The activities include the following:

1. Fence intruder detection cable;
2. Electrical heat tracing panel;
3. Electrical heat tracing cable;
4. Tray work;
5. Cabling;
6. Cable termination;
7. Conduit and perforated tray;
8. Grounding work;
9. Fire alarm work;
10. Power field equipment, lighting, and receptacles.

And these are shown under seven respective areas in the Level 3 Schedule.

Apart from seven areas, Electrical and Cabling work tasks are substantial in buildings, especially in substation building and technical building. The tasks are therefore shown for every building in the Level 3 Schedule. In substation, panel installation, transformer installation, conduit, and perforated tray installation are additional tasks.

In technical building, lighting and receptacle, fire alarm detector, tray work (for cellar) are additional tasks. Apart from technical building, there are several other buildings, and each building needs lighting, and hence, related activities are to be reflected per building.

7.2.9 HSE

HSE Manager, the discipline head, stationed at Site reports to both Site Manager and corporate HSE Head at headquarters. HSE Management is done as per HSE Plan and Procedures. Weekly meeting with Owner takes place to review the performance and compliance. Design HSE, being part of engineering discipline, is different from HSE. Contractor's HSE Management comes to an end when Owner team takes over the Plant upon hydrocarbon feedstock introduction or at the issuance of PAC. During Plant start-up and commissioning, HSE Management has the responsibility to operate firefighting systems (deluge of water and foams) to handle emergency situations.

The day-to-day HSE activity at construction Site, among others, is as follows:

- Check the health certificate of any construction equipment (mainly crawler cranes) upon mobilization to Site and prior to allowing them for use in construction;
- Check scaffoldings for its safe installation and installing red or green tags as appropriate;
- Check if workers and supervisors wear safety helmet, harness, shoes, goggles, etc. Check on workers who work at heights;
- Check grounding of electrical welding and other equipment so that electrical shocks are eliminated;
- Conduct every morning tool box and demonstration of safety standards;
- Check on house keeping;
- Check if hazardous materials are safely kept and necessary barricades are provided;
- Check if confined space working permits are taken and work is monitored;
- Check if all the moving vehicles have spark arrestors;
- Check if electrical power distribution boards are proper and live bare wires are disallowed;
- Monitor how waste disposal are taking place;
- Monitor if temporary sewage systems are working properly;
- Monitor diesel and gasoline storages are properly secured and protected from fire hazards;
- Check if all excavated and open pits are barricaded and illuminated with lighting;
- Check if proper slopes are provided when deepening of pit takes place during excavation;
- Conduct drills for fire hazard and emergency evacuation responses;
- Check how emergency alarms are generated and responded;
- Check readiness of ambulances and medical, Medevac;
- Provide inputs for daily reports under HSE section on the construction to field control discipline;
- Update organization chart with pictures of office bearers;
- Provide daily display of photos on wall-mounted boards for making awareness;
- Monitor Speed Control of Vehicles: Display speed limit stickers glued to inner of the windscreen ahead of driver seat.

Notwithstanding the above, the tasks to be indicated in Level 3 Schedule are only the following:

- HSE Management duration at Site;
- Installation of firefighting systems (reflected for each area and buildings).

7.2.10 HVAC

The activities include typically the following:

- Installation of equipment and materials such as AHU (air handling unit), condensing unit;
- Ducting;
- HVAC panels;
- Cabling;
- Piping/plumbing.

The buildings for which HVAC equipment is installed include control room, laboratory, technical building, and office building. The above-said bulleted activities repeat under each building for Level 3 activities schedule.

As said earlier under engineering and procurement section, HVAC systems are designed and supplied as a package by Vendor. Contractor may have the discretion to have the installation part executed by itself under the supervision of Vendor. Copper tubing is one activity at Site that may assume importance in this area during construction.

HVAC units are installed adjacent to the building into which the ducting enters to provide air conditioning.

7.2.11 Painting, Insulation, and Cathodic Protection

The following are the typical tasks in Level 3 Schedule, and they appear for each area wherever applicable:

- Field painting of piping systems and others;
- Cathodic protection (cathodic protection requires boring deep into the ground/soil to place electrodes and wire);
- Insulation of equipment;
- Insulation of piping (hot and cold type).

The equipment are shop-painted, and hence, there is no need to paint in field except touch-up or special cases. Insulation works of hot and cold type are performed at Site after completion of piping systems' hydrotesting and reinstatement.

7.2.12 Telecommunication

The activities are installation of field devices, system cabinets, and panels in various buildings and cabling.

Telecommunication system is an activity identified for each area. It may include CCTV, radios, and other telephonic systems with satellite towers if needed. Telecom tower is installed at office building.

7.2.13 ICSS

Installation of the ICSS panels in the control room building and cabling to field instruments/panels.

7.2.14 Water Wells

This activity may assume importance if the project Site is situated in a desert where water is taken through bore well pumps, which require boring to a depth as deeper as 1 or 2 km. At the beginning, the usage can be for temporary facility and labor camp operation, later for the operation of the Plant.

7.2.15 Miscellaneous

Keep this section to address some miscellaneous items in the Level 3 Schedule.

7.3 OFF-SITE CONSTRUCTION

As the project of upstream dealt in this book is assumed to have some wells, well pads, flow lines, and trunk lines, these facilities altogether are termed as Off-Site, in order to distinguish it from gas treatment/processing facilities, which is called Main Plant.

The Level 3 activities in Off-Site constructions are shown here below.

7.3.1 Site Preparation

Site preparation activity addressed here is related to well pads and clusters, which are far away from central processing facility. Site preparation of well pads is recommended to be done after the handing over of wells by Owner. Handover means completion of drilling of the well and rigless test. Handover of well pads before completion of rigless test is not a recommended practice. Otherwise, to perform the rigless test of the well, Owner may have to suspend ongoing Contractor's activities in this vicinity citing safety grounds. In such case, Contractor may suffer standby cost and time loss.

Contractor can explore innovative ideas (through Value Engineering) during site preparation to reduce the volume of work by raising finished floor level to cut cost. Such measures are useful only when the subsoil conditions are rocky or hard.

7.3.2 Foundation works

Well Pads: Works in this area include foundations or pedestals for pipes, pipelines, solar panels (in case of desert site situation), hydraulic power pack units, fence, manholes, and valve pits.

Cluster Area: In addition to what were stated under well pad, foundations are required for firewalls, pig launcher, and pig receivers.

Trunk Lines and Flow Lines: Civil works under this are predominantly excavations, and such excavations could be very tedious if the subsoil contains rock. Heavy-duty rock trenchers are used to excavate the trench for pipelines if hard rock is seen in subsurface. Heavy-duty rock trenching is expensive, and these machines can be used when rock blasting with explosives is not permitted.

7.3.3 Structure Works

Structure works are simple in nature, and these are platforms for well pads and clusters.

7.3.4 Equipment Installation

The equipment for clusters are pig launcher and receiver, HP cold vents, methanol injection package, corrosion inhibitor package, and mobile pig launchers. The installation of each above item is an activity.

7.3.5 Pipelines Installation

For well pad, the activities include above-ground piping fabrication, pipe support fabrication, large bore and small bore piping erection, pipe support erection; for trunk lines and flow lines, underground piping installation at field.

7.3.6 Roads

Fence and gates, gravel and concrete paving, asphalt roads (along trunk line):

If these activities are found in clusters and well-pads areas, suggestion is made to show individually these activities under such areas for scheduling purpose.

7.3.7 Field Instrument and Cabling

The activities include cable tray/conduit installation, multicable laying, cable termination, secondary cable laying, junction and field box installation, panel installation, instrument calibration, instrument installation, air pipe, and SS tubing.

7.3.8 Electrical and Cabling

For Well Pads and Cluster: The activities include tray and conduit work installation, cable laying, cable termination, earth grounding, panel/cabinet/junction box installation, and field items installation.

For Trunk Lines and Flow Lines: underground fiber optic cable installation.

7.3.9 HSE

This work addresses how health, safety, and environment are managed at Site by HSE team who may be present at central processing facility while occasionally at Off-Site depending on the exigency.

7.3.10 HVAC

No HVAC work is involved under HVAC as no buildings are assumed to exist in Off-Site. Hence, there is no need to show this under Level 3 Scheduling.

7.3.11 Painting, Insulation, and Cathodic Protection

Cathodic protection, field painting, field insulation, and this cathodic protection are for protecting buried facilities from galvanic corrosion.

7.3.12 Telecommunications

Telecom Tower, Field Devices for Each Well Pad and Cluster: Activities are identified and stated.

7.3.13 ICSS

Off-Site ICSS Work: No ICSS activity is involved.

7.3.14 Miscellaneous

Off-Site miscellaneous work.

8 Pre-commissioning

8.1 PRE-COMMISSIONING OF THE MAIN PLANT

Pre-commissioning activities are executed and monitored system-wise and subsystem-wise. In construction phase, activities are monitored area-wise. As such, the monitoring approach for pre-commissioning obviously varies from Construction. Even when systems and subsystems are defined, still there is a possibility to reflect them under area. Pre-commissioning is successor activity to Construction. In Level 3 scheduling, pre-commissioning activities appear under system and subsystems and also simultaneously under respective area. System definition is an important task, and this is prepared by Contractor and approved by Owner at least 6 months prior to start of pre-commissioning. System-wise approach is usually prescribed for pre-commissioning for most Contracts. Further, software tools are available for monitoring pre-commissioning activities. In the Detailed Level 3 Schedule, such system-wise schedule need not be incorporated in the beginning. However, when the Construction work progress reaches 50%, it may become necessary to have a Detailed Level 3 Schedule that addresses the pre-commissioning and commissioning activities system-wise. Fifty percent Construction completion heralds the opening of work fronts for pre-commissioning.

8.1.1 General

Deluge/Drains Systems

Continuity checks, final alignment, and vessel gross leak tests are activities applicable to deluge, closed and open drain systems. When deluge system opens to operate, water jets discharge speedily at high speed from nozzles of a ring header pipes that surround the equipment. The equipment is cooled quickly to prevent accidental explosion. Drain systems are to ensure that all the liquids discharged are taken into the desired location as quickly as possible.

8.1.2 Utility Facilities

Under the utility, the subsystems may appear for each of the fire water, source water, fresh water, potable water, sewage, waste water treatment, air, nitrogen, and power generation. The pre-commissioning tasks are again:

- Continuity test (electrical and instrument);
- Final alignment;
- Vessel box up;
- Vessel gross leak test;
- Hydrotest on above-ground piping where applicable;

- Flushing ad blowing;
- Reinstatement of instrument valves that have been removed to facilitate hydrotest/flushing and blowing;
- Lube oil flushing for power generation unit where turbo-generators are installed.

8.1.3 Pre-commissioning of Main Plant Process Systems

Under each area, and under each subsystem, the pre-commissioning activities as aforesaid can be listed, executed, monitored, and controlled. In some cases, a group of few subsystems may run under more than one area to serve a common purpose, and they are related to utility and fire-fighting systems. Hence, they are addressed separately at the start.

8.2 PRE-COMMISSIONING OF THE OFF-SITE

8.2.1 Pre-commissioning of the Off-site Utilities

Fiber optic cable test, hydrotest, flushing and blowing, and pig cleaning are typical activities, which may come under clusters and under well pads.

8.2.2 Pre-commissioning of the Off-site Process Systems

There are no activities.

9 Commissioning, Start-up, and Performance Test

9.1 COMMISSIONING ACTIVITIES

Commissioning is a successor activity to pre-commissioning.

Typical commissioning activities are as follows:

- Solo run (of motors or drives specifically);
- Loop and function test;
- Operation leak test;
- Hot alignment (hot alignment is tightening of bolts in hot condition to ensure that the alignment of compressor, turbines, and rotary machinery is properly secured to its foundation, and at critical joints by inducing the required tension into the bolt, hot alignment is also applied to flange joints of higher rating and critical service piping; equivalent tasks apply where such bolting is absent);
- Operation test;
- Drying and inerting;
- Chemical loading (scrubbing solvent for acid gas removal system);
- Hot oil filling;
- Putting into service the utility systems such as air and nitrogen.

During the commissioning of the Plant, the local authority inspectors visit and check if all fire protection and fighting systems are completed in all respects for handling any emergencies at the start-up. They also check to ensure that temporary scaffolding works are removed, and the escape routes/passages around equipment, permanent ladder, and stairs are free of obstacles. Authorities grant permission to Contractor for hydrocarbon or feedstock intake into the Plant only after its inspectors have inspected the Plant and provided a favorable report.

Authorities also may monitor the unnecessary waste of hydrocarbon that may result in prolonged flaring caused by process upsets occurring between start-up and re-start-up. If the flaring quantity allowed in the issued permit has exceeded, further permission is obtained for increased forecast quantity in flaring of gas into atmosphere.

As Owner's operators should be in a state of readiness to participate with Contractor's team in handling the commissioning and start-up and operation, the operator's training should have been completed by this time.

Further spare parts should also be available to meet exigencies and replacement of parts.

9.1.1 Commissioning of the Central Processing Facilities

The above-mentioned commissioning tasks of subsystems falling under its respective area can be executed and monitored. However, as said under pre-commissioning, some subsystems such as utility subsystems are not restricted to one particular area but are designed to spread and run across other areas. Such spread is unavoidable due to the network of fire water line required under safety considerations. These networks extend to cover all vital areas to fight fire hazard that may occur during operational upsets and maloperation of the Plant to any equipment.

9.1.2 Commissioning of the Off-site

As Off-Site covers only the well pads, clusters, trunk lines, and flow lines, the commissioning tasks are limited to a few loop tests, operational leak tests, and inerting.

9.1.3 General Precautions for Certain Items

Interlocks

The critical units in this type of Plant are acid gas regeneration unit and gas turbine generators (related to power production or compression of process fluid). Often the interface logic between turbine manufacturers and compressors could be an issue during start-up, and such engineering issues are to be tackled. A dedicated/joint session in the engineering phase to align the logic is required.

Antifoaming Injection

The commissioning of AGRU is very challenging task as aMDEA can create foaming if pressure and temperature are not properly controlled within stripper and CO_2 absorber during the start-up.

During the start-up, if foaming appears in uncontrollable manner, antifoam injection should be activated to see an improved result. There are two types of injection: one is to take care of emergency situation/process upsets where more volume of injection is required, and the second is normal injection which operates through lesser-capacity pumps. If this intervention doesn't help, allow the liquid to rise until HHL to trip and isolate the AGRU. Another alternative is to reduce the load into AGRU by reducing the flow, and this can be achieved by using the line meant for blow down off feed gas coalescer by activating through ESD, and this could trip the AGRU and isolate it, while MP compressor at its upstream can be allowed to run with gas in recycle mode. Stopping MP compressor should be the last option as re-start-up is not easy. If the PSV bypass line is used to vent out the gas, which is violation of design, the line may tear off from vent line. This bypass valve is generally through ball valves of size two inch diameter with NC (normally closed), and this arrangement of bypass is designed for maintenance purpose. Hence, it is necessary to give an orientation on process venting/draining by process team to commissioning team.

9.2 START-UP ACTIVITIES

The start-up is for the whole PLANT. Prior to the start-up, the following systems should have been in operating condition and normal operation:

- Fire water systems;
- Deluge systems;
- Fire protection systems;
- Air and nitrogen service systems;
- Power systems;
- Water utilities systems;
- Cathodic protection systems.

The start-up sequence varies from Plant to Plant, and the following start-up sequence is typical to central processing unit (CPU) with well pads and cluster shown in a process block flow diagram shown in Appendix 6.

1. Line up and service first well and its related cluster;
2. Feed in and gas reception at the first unit of CPU;
3. Feed in service and service HP/LP flare;
4. Firing and activation of HP/LP flare;
5. Power normal and service load up;
6. Circulation for hot oil system;
7. TEG (glycol) filling and circulation for dew pointing;
8. Amine filling and circulation for acid gas removal system.

Although control loops' overrides could help stabilize the Plant quickly at the start-up, overrides should be jointly discussed and agreed between Owner and Contractor in order to ensure safe Plant shutdown when abnormal operational upsets take place. Even if the Plant at the start-up faces several trips due to pressure and temperature variations, Contractor's request to have overrides on control systems can be approved provided parties have performed safety reviews for the requested overrides. When overrides are minimized at the start-up for safety reasons, the frequency of Plant trips at start-up could be frustrating, yet safer.

9.3 PERFORMANCE TEST

The basic tasks in Level 3 Schedule are as follows:

- Gas feed in process;
- Load up and stabilization of various units with minimal overrides on control system;

- Function test;
- Stable and normal operation 96 hours (24 and 72 hours);
- Performance test runs.

Apart from performance test said above, a Summer Test is additionally recommended for those Plants which are constructed in countries where extreme hot weather with a significant temperature variation across the year is experienced. The operational efficiency of air-fin coolers when ambient temperature jumps to higher level in summer is to be measured in a Summer Test, if the performance tests are carried out during winter.

There are Licensor and Vendor packages requiring validation of process parameters and guarantee figures, for which testing, their representative's witness during testing are required. If process gas is to be exported from battery limit (at fiscal metering unit) to other party's export line, then both parties' agreement together with authorities' approval on calibration is required. Even a very negligible or unnoticeable error in calibration can cause huge financial losses when the error is accounted for the whole year.

10 Temporary Facilities

Temporary facility (TF) consists of office facility for construction phase, and boarding and lodging for direct and indirect workers/staff. It is designed, built, operated, and maintained by Contractor. The task of TF is included in the scope of work of Contractor, and in such case, design and engineering, procurement, and construction are detailed out. As this facility is to be established in the beginning, care must be taken not to delay these activities. If delayed, the mobilization of workers and supervisors will be delayed consequently. Facility may include camps for Owner, Contractor, and Subcontractor. The facility may extend to include military personnel if security is an issue and to be implemented as per government directives. The establishment of TF is very important but is often delayed in most of the projects. Owner should have these facilities designed to the degree of Front End Engineering Design through parties other than Contractor prior to award of Contract or commencement date. Otherwise, design of TF could take as much time as does the design of permanent facility/establishment when Owner applies a common design review and approval criteria for both cases. It is advisable to have this kind of facility no later than ED Plus 2M. Design for sewage treatment, bath water piping, electrical power supply, internet, and communication network could take considerable time. ED corresponds to effective date of Contract from which the project timeline starts.

If the Contract stipulates a design basis and scope specifying the capacity, size, and number of personnel to be housed, with no Front End Engineering Design made available thereof, then Contractor takes up the Detailed Engineering Design from the effective date. If Owner alters the design basis or scope, it could put both parties into change order approval process and delays. Hence, it is advisable not to alter the design basis or scope for TF. Delayed completion of TF prevents the mobilization of manpower and resources. The delays to mobilization of manpower and resources impact adversely on the Contract deadlines and project duration of the permanent facility/Plant.

A separate Level 3 Schedule is prepared and maintained to perform and monitor this work. And this schedule is an integral part of Early Work schedule, which is different from Level 3 EPC Schedule for permanent facility.

Since weight factor is allocated for this work and price breakdown shown in the Contract price, it becomes necessary to have progress of this work measured every month and certified by Owner in order to enable Contractor to raise invoice for this portion of the work.

The TF is designed to include the following features:

- Residential accommodation for Owner's staff during project phase;
- Residential accommodation for Contractor's staff during project phase;

- Residential accommodation for Subcontractor's staff: each Subcontractor may have his accommodation separated from other by fence to avoid HR issues;
- Residential accommodation for security staff during project phase;
- Messing and recreational facilities for each party as said above;
- Parking places and space open and closed door sports;
- Prayer rooms;
- Library and gyms;
- Office building for Owner;
- Office building for Contractor;
- Office building for Subcontractors;
- Warehouse and storage;
- Batching plants;
- Fabrication yard;
- Laydown area;
- Backfilling area;
- Security cabins and post;
- Watch towers for vigilance at vantage points along boundary;
- Medic and clinic.

Section II

Commercial Aspects

In this section, there are 22 chapters on the following topics:

Commercial date and effective date; extension of time; professionally handling large claims; prolongation and disruption costs; penalty/interest for payment delays; force majeure; concurrent delays; hedging against currency fluctuation; indemnity; consequential, indirect damages; law and regulations; escalation; liens; total liability; payment terms; planning; progress measurement; EOT claim; interpretation rules; change/variation/trend notices; contract price below floor price; owner's delay on document approval; and insurances.

11 Commencement Date and Effective Date

Commencement date or effective date is the most important key date. In schedule preparation, this date is called zero date. Owner generally requires the project to commence as early as possible, and it accordingly attempts to fix an effective date to suit its requirement but Contractor, in view of the following, should exercise caution before accepting a date earlier than what could be reasonably possible.

The following activities have to finish before zero date.

1. If both parties, at the conclusion of technical clarifications during the bidding stage, understand there is a possibility for the addition of new units, feed gas compositions change, pressure drop revisions across the process line or units relocation and a consequent need to modify the FEED particularly in respect of HMB, PFD, and P&ID or process data sheets, they should perform time impact analysis jointly in the critical path of the schedule to determine the overall impact on project duration. If additional time is foreseen, then necessary adjustment is to be made in the Contract with respect to project duration and Contract price. Instead, if parties desire to settle this issue during project execution, then they have chosen to undergo a turbulent journey and complexities.

2. If the Contractor is a temporary association of Joint Venture (JV) members, there is a need to legally form an entity in the country where the Site is located. Formation of entity may take as much as 6 months' time due to drafting and finalization of the following documents:

   - JV execution agreement;
   - Shareholders agreement;
   - Memorandum of articles.

   Unless the legal entity is formed, payment to members of the JV cannot be realized from the payments made by Owner. Hence, for 6–7 months, the cash-in will be zero, and so each partner has to make payments on their own for the following works:

   - Site preparation;
   - Soil investigation;
   - Building temporary facility;
   - Salaries and travel expenses of employees;

- Rental of office;
- Utility consumption and administrative expenses.

3. General terms and conditions for PO and Subcontract should have been formalized prior to zero date to avoid a delay in mobilization of soil investigation Subcontractor, if the soil data were not to be relied upon.
4. Construction All Risk and Third-Party Liability Insurance are procured quickly to facilitate the mobilization of Subcontractors.
5. No work Visa can be issued for personnel without the formation of legal entity of JV partners, and hence, this could impact the movement of personnel to a country where engineering would take place or to a country where the Site is to be opened.

12 Extension of Time

Time extension means the time extended beyond the Contract Deadlines until a date such Contract Deadlines are re-set by Owner. Normally, Contract Deadlines refer to the following major milestones. Liquidated damages are slapped at these Contract Deadlines.

- Mechanical completion;
- Ready for start-up;
- Provisional acceptance certificate.

When Contractor claims time extension, it is incumbent upon it to prove through Schedule analysis how events have delayed the Project, and how Contractor caused delay events are not concurrent with Owner caused delay. As per Industry Practice, prolongation cost corresponding to the granted time extension is disallowed if concurrency of both parties' delays is present.

13 Professionally Handling Large Claims

For a professional treatment of a large claim, Contractor approaches professional organization for analysis and report. Owner has natural tendency to respect claims prepared through third-party professional organizations, and it is difficult for Owner to reject claims so prepared. For this purpose, Contractor should maintain without fail monthly schedule reports with analysis on delay events especially on critical path. The Planning Engineer needs to preserve the approved schedule without updates, which is called un-impacted schedule. In order to project any single delay event into the schedule to see if the Contract Deadlines are delayed on account of one such delay, the un-impacted schedule is used. Similarly, all the delay events, one by one, are to be projected into the un-impacted schedule to see the impact in the critical path. Un-impacted schedule means the schedule which does not present any delay of any activity. Written notices served to Owner on delays, including the ones below, will add value and help the professional organization to analyze the impact on critical path:

1. Document review time more than allowed in Contract;
2. Unjustified rejection documents of Category 1 as listed in Appendix 7, where no float is available;
3. Site access is delayed due to handover of areas where Owner delays are accounted;
4. Owner wishing to have bigger changes executed having impact on time;
5. Feedstock availability delays;
6. Poor quality of feedstock that creates obstacles in stabilizing Plant before conducting performance tests;
7. Exporting lines of third party not ready to receive product gas from the commissioned Plant;
8. Delays in well drilling, which may prevent handover of sufficient number of wells and quantity of gas for Plant load up and stabilization;
9. Change in law impacting on critical path, especially restriction on security-sensitive communication instruments;
10. Owner consent delays caused disruptions;
11. Delays arising from authorities in the country where project is executed, despite due diligence exercised by Contractor;
12. Delays arising from encountering, during excavation, subsurface obstructions unforeseen in the Contract;
13. Encountering fossils or monuments or religious relics in excavations;

14. Suspension of work by Owner due to financial difficulties for some time (normally this is usually a undisputable item as it is at the request of Owner or the decision of Contractor using Contract clause under nonpayment scenario or due to other reasons.)

Any delays should be recorded by Contractor in official records (letters, not emails as emails are not recognized as official communication channel) from time to time expressly stating the reservation of right to entitlement in the future and such letters shall have subject/caption: Notice of Delays-Delay Event xxx.

Any such letters written shall be free of false accusations, impolite and unethical languages.

Contractor should also have some understanding of what excusable and non-excusable delays are and what compensable and non-compensable delays are.

Compensable means entitlement to prolongation and/or disruption costs in addition to grant of time extension is qualified/justified.

Depending on the gravity of the situation and the degree of concerns, letters served above should bear captions "Without Prejudice".

14 Prolongation and Disruption Costs

When the Contract Deadlines (MC, RFSU, and PAC) are delayed due to various reasons, mainly for reasons attributable to Owner, Contractor should submit an EOT Claim. Such claims are supported by prolongation cost and disruption cost.

14.1 DISRUPTION COST

If Contractor's planned work process is disrupted due to actions of Owner, Contractor may suffer productivity, consequently incurring higher costs of execution. For example, P&ID issuance and approval are critical items. As start of activities of engineering of many disciplines depend on this P&ID, the rejection of P&ID by Owner could mean brakes on the forward movement of critical activity, consequently delaying the Contract Deadlines. It could be considered as disruption. On the whole, it is the Contractor who is responsible for completion on time and start of commercial production. Owner should understand such situation and approve the P&ID with comments instead of rejecting it. Owner may even ask for additional issue of P&ID if it desires to ensure a better quality, and in this manner, it can avoid rejecting a P&ID.

Another case of disruption can surface if Owner demands multiple times submission of procedures/drawings merely for the sake of making comments. Normally, engineers who are entrusted with task of this approval from Owner's end focus more on quality of the document than on the consequences to the critical path of schedule. Contractor also may find in many cases, the review time taken by Owner exceeding seven or ten working days before returning the deliverables to Contractor. Such situation also disrupts the planned working of Contractor. In order to claim the disruption cost, Contractor should notify standby time thereof, or at least summarize weekly basis a time sheet indicating how man-hours in engineering have been underutilized in comparison with the budgeted man-hours. Budgeted man-hours are to be shown in the Level 3 Base Line Schedule in terms of histograms that Contractor submitted at the time of bidding and then validated it as Contract-based figures.

See Appendix 9 for the computation of disruption costs.

14.2 PROLONGATION COST

Prolongation cost is for the cost of prolonging the indirect manpower and establishment. In prolongation cost, the direct manpower costs are not accounted for, nor payable. However, if direct manpower costs are included due to interference or disruptive action of Owner, then such cost of standby can be claimed under disruption costs.

For instance, if the project construction is delayed 10 months for reasons not attributable to Contractor culminating from the delayed approval of foundation design drawing, access to Site, or engineering in general, then prolongation cost for the 10 months can be included in the claim. There are two ways to do it: one way is to compute the actual average man-months on the indirect manpower and other establishments (temporary facilities/camps/etc), and second, by following the Hudson formula or alike. The computation method is explained in Appendix 9.

15 Penalty/Interest for Payment Delays

Cash flow is regarded as the backbone of the project. If the history is looked into, there are instances of termination notices served by Contractor for reasons of delayed payment or payment default by Owner, and such actions had been upheld in arbitration awards.

If Owner delays the payment that corresponds to monthly progress, Contractor can hold Owner liable for the delay and demand interest payment for the period delayed. Contractor can ask 3%–4% above 6-month LIBOR. If LIBOR appears to be 2.5% for the 6-month average, the total interest per year would stand at $4\% + 2.5\% = 6.5\%$. LIBOR is London Interbank Offered Rate for lending/borrowing between banks. LIBOR varies according to market, and hence, it is not a fixed value. Borrowing a working capital from banks will be at LIBOR rates; besides, there are processing fees and other fees, which all sum up to 6.5%, the cost of borrowing.

16 Force Majeure

Whenever a part or whole of work is affected for reasons attributable to unforeseen natural phenomenon or other, Contractor has a tendency to write to Owner terming such incident as a Force Majeure event. Even when definition of Force Majeure as provided in the Contract may not qualify any particular event on which Contractor eyes to obtain relief from Force Majeure, Contractor makes every attempt to seek protection under this Force Majeure for a part or whole of work impacted by the event in question on which he has no control or for which he is not the cause. Contract clause is normally proffered by Owner. Owner while drafting and awarding the Contract tries his best to deny any room for Contractor to bring up any excuse. During execution of work, a timely notice of such event and the duration of work stoppage should be officially recorded and notified by Contractor, and such action could help Contractor to obtain time extension and commercial benefits at the completion of the project.

Force Majeure events are unforeseeable and cannot be factored in the Project Timeline or Schedule at the time of entering into the Contract. As far as Contractor is concerned, Force Majeure event arises when it is rendered unable to perform the work in part or whole under a Force Majeure event over which it has no control. If the Force Majeure situation prolongs longer time, suspension of work could be an appropriate action in terms of Contract and a notice to this effect should come from Owner as per provisions of suspension clause therein. Some example of such event could be as follows:

- Act of civil or military authority;
- War;
- Fire or natural calamities such as earthquake, deluge, sand storms, Tsunami or abnormal heavy rains and floods;
- Strikes, lockouts, if considered under the law, provided the cause of such event is not arising from the mismanagement of the Contractor;
- Civil disturbances;
- Explosions;
- Terrorism;
- Outbreak of pandemic diseases.

17 Concurrent Delays

If the reason for delay of a Milestone or Critical Activity is attributed to both Owner and Contractor, such delay can be treated as concurrent delay.

Let us consider the following cases A and B where both parties caused delays:

- Case A: Owner has taken more time than allowed to complete his task of review on Foundation design drawings (IFC, issued for construction) for a critical equipment Compressor; such delay, for instance, is counted as 20 days, which falls within the timeline of 45-day delay caused by Contractor as below.
- Case B: Contractor is unable to make its batching plant ready to produce the concrete and hence is not in a ready position, but such delay runs concurrent to delays in Case A above. A delay caused by Contractor here was 45 days.

Although 45 days has impacted the critical path and the PAC (provisional acceptance certificate), Contractor can claim only 20 days for time extension without penalty.

18 Hedging against Currency Fluctuations

Hedging provides the Contractor with a buffer from shock and ability to mitigate the foreign exchange risk. The currency risk arises when downslide in currency happens at the time of Contractor's making payment to his Vendors/Suppliers in a currency different from the currency to be received from Owner.

In principle, payments are to be made to Vendor by Contractor from a source of receivable that comes from Owner during execution of the project, but the payments from Owner to Contractor may not come at the same time as Contractor are to pay to the Vendor.

Let us assume that the currency exchange rates vary in money market. If the currencies of contract and purchase order (PO) are the same, then the risk on currency is fully eliminated, but in most of the projects, the currencies are different. "Contract" is between Owner and Contractor, while "PO" is between Contractor and Vendor.

To avoid any unforeseen financial loss arising from currency fall, Contractor can have the choice of entering into a forward contract with a hedging company at the award of Contract, by which it can lock in an exchange rate to future dates at which he expects the currency transaction should happen during the span of the project execution.

The exposure to risk is to be assessed. Hedging of the Contract-based currency is to be initiated at the commencement date of the Contract, and this is primarily the responsibility of the Project Management Team of Contractor.

19 Indemnity

As a standard practice, Contractor is required to indemnify, hold harmless, and defend Owner and its representatives/officers, agents, and employees from and against claims from third parties or governments related to

I. Noncompliance of law, regulations;
II. Infringement of know-how, patents, copyrights;
III. Willful act or gross negligence causing death/injury/property damage;
IV. Property damage/injury/death (if not caused by gross negligence or willful act);
V. Public nuisance;
VI. Pollution, contamination.

For the items IV, V, and VI, insurance policies can cover, but for items I through III, cannot.

Since the standard clause in every Contract on indemnity is elaborate, and its language difficult for an untrained Contract Engineer to understand, the summarized form is given above.

20 Consequential, Indirect Damages

The term 'consequential damages' is often misunderstood or misinterpreted. UNCITRAL has termed consequential or indirect losses to be the losses causally remote from failure to perform. The foreseeability factor is very important for recovering the consequential loss.

In simple expression, a Contractor is not liable to a Vendor for consequential and indirect damages; similarly, a Vendor is not liable to a Contractor.

Let us look into the following phrase of a contract article, which is prevalent in many contracts, where the Purchaser refers to the Contractor and the Vendor to those who supply material and services to the Contractor in a Project. Let us further assume that the aggrieved party is the Purchaser as a result of the actions of the Vendor.

> In no event shall the Vendor be liable for any special, consequential, indirect or incidental damages, including, but not limited to, loss of anticipated profits or loss of use of any equipment, installation, system, operation or service into which the materials may be put or services performed. This limitation on Vendor's liability shall apply to any liability for the default under or in connection with the materials delivered hereunder, whether based on warranty, failure of or delay in delivery or otherwise.

Most legal systems have rules for determining losses of consequential and indirect damages.

The parties may therefore wish to leave this issue to be resolved under the rules of the law applicable to the Purchase Order. In addition, if the parties have identified specific kinds of losses, which they wish to exclude from compensation because they would be too remote, the parties may so provide.

For example, the following losses cannot be treated as either indirect or consequential loss suffered by a Contractor/Purchaser:

- A diminution in value of assets of the Purchaser not supplied by the Vendor, e.g., damage to property of the Purchaser not supplied by the Vendor resulting from defects in equipment supplied or construction effected by the Vendor; (According to Anson's law of contract, the assessment of Purchaser's/aggrieved party's loss of bargain will be the difference in value between the performance received and that promised in the Purchase Order; "diminution in value".)
- Expenses reasonably incurred by the Purchaser/ aggrieved party which would not have been incurred if the Vendor had performed his obligations, e.g., wages paid to personnel hired by the Purchaser to commence construction during a period in which construction cannot be commenced because of a failure by the Vendor to deliver the equipment and material;

- Payment which the aggrieved party/Purchaser makes to a third person because of liability to make those payments arising as a result of a failure to perform by the Vendor; and
- Loss of profits, which would have accrued to the aggrieved party/Purchaser if the Purchase Order has been properly performed by the Vendor.

Clarity on "indirect damages" will arise if its opposite "direct damages" is first understood.

The following explanations are available from the ENR Fall Conference Report 1980.

Direct damages include

- Wage and material escalations;
- Excess and extended general conditions;
- Extended equipment use, and;
- Extras generally.

Consequential damages include

- Interest in interest rates that occurred during a delay. It was held that increased rates were beyond the control of the parties and would not be considered as part of the damages claim unless it could be shown that parties have foreseen this expense at the time of contracting.

Following is a popular text for phrasing the terms and conditions related to this article from the American Institute of Architects (AIA):

> Waiver of Consequential Damages by Contractor/Purchaser and Vendor:
> Both Purchaser and Vendor agree to waive all claims against each other for consequential damages.

Accordingly, if there is a breach of Purchase Order resulting in a delay, the Purchaser agrees to waive such claims as additional rental expenses, loss of income or profit, and loss of management and employee productivity resulting from breach. Vendors on the other hand agree to waive claims for expenses such as office expenses, additional compensation for personnel stationed at offices, and damages resulting from loss of financing, business, and reputations, and loss of certain profits. The Purchase Order does allow the parties to mutually agree to the award of "liquidated direct damages" resulting from the delay.

Let us consider the following Purchase Order clauses observed in Projects to further our understanding of the gravity and umbrage of phrases like "arising out of" while drafting clause under consequential damages.

- Unless stipulated in the Purchase Order, Vendor shall not be liable for any claims, loss, or liability *arising out of* or *in connection with* the Purchase Order for indirect or consequential damage;
- Vendor shall not be liable to Purchaser for any special, indirect, or consequential damages such as a loss of profit, loss of production, or loss of

> sales contract arising out of or in connection with or resulting from this Purchase Order, whether or not such damages arise out of or result from the negligence of Vendor, its employees or agents. With the exception of liability for special, indirect, or consequential damages arising out of or in connection with or resulting from a violation of provisions of Article on Secrecy, Vendor shall not be liable to Purchaser for any special, indirect or consequential damages such as loss of profit, loss of production or loss of sales contract *arising out of* or in connection with or resulting from the negligence of Vendor, its employees or agents except to the extent such damages are covered by insurance to be procured or maintained by Vendor under this Purchase Order, and except as otherwise described in this Purchase Order.

In the above, one will see phrases such as "arise out of," "in connection with," and "resulting from." English courts have given the widest meaning to the phrase "arise out of," and not to others. And this form of words will usually embrace all disputes capable of being submitted to arbitration. The mere use of words "under this contract" will be interpreted as excluding any claims other than pure contractual ones, as authors Martin Hunter and Alan Redfern state in their book *Law and Practice of Arbitration.*

21 Laws and Regulations

Each Contract stipulates the law of contract to be the law of a country where the contract is executed. A Contractor if foreign based with reference to the country where the Site is situated may prefer a neutral country's law to be the law of contract. However, the English law is widely popular.

Most of the Nations do not have a comprehensive law to so clearly address all issues as does the UN organization for example in the CISG law.

For example, the CISG law of UN can apply to a Purchase Order, which is placed by a Malaysian Contractor on a US-based Vendor for a supply of equipment to be delivered to a Site in Thailand.

According to the CISG law, there are several legal aspects and guidelines available to find a solution. As most of the Nations have already accepted such laws through access/ratification/etc. as can be seen from the Scoreboard of Adherence from the official site of UN, the CISG law fills normally the vacuum in each Nation's law when a transaction in trade involves the international sale of goods.

An example is a turnkey EPCC contract. A single turnkey contract may contain a primary part 75% offshore (costs wise) work that falls into the territory of many countries and a secondary part 25% field construction to be executed in the country where the Site is located. The Site is assumed to be in Thailand. For the secondary part, the contract law applicable should be the law of the country where the work is executed, whilst for the primary portion, it should be a different case. The ratio of 75:25 can vary from Project to Project.

Let us assume that in the purchase of equipments falling under the primary portion, there is a gas compressor whose driver is a gas turbine. The Contractor is, for example, of Malaysian origin who is to purchase this compressor together with the driver turbine from the manufacturer of US origin and also other equipment under the primary portion from manufacturers of other countries.

If the equipment is manufactured in one country (here in the USA), then naturally the law of that country (USA) should be followed, because there is some public interest corresponding to each law, for example, restrictions on foreign exchange, import, export, tax duties, environment, and construction safety during manufacturing. Hence, often it is impractical to assign a local law of Thailand where the Site is situated for procurement of compressors and turbines, which are being manufactured in the USA.

In another example, for an EPCC contract between Contractor (Malaysian based) and Owner (Thailand based), the choice of the law selected for the contract is the law in force of the Kingdom of Thailand, which in turn refers to the UN law, so is the case with many other countries.

22 Escalation

The Contract Price sometimes can be significantly affected due to a sharp or abnormal increase in the Price Index caused by an increase in oil and gas prices. If wage and steel prices driven by the oil price increase every year, this increase could significantly reduce the anticipated profit. The delayed period of any Project encounters this escalation scenario. Several mega projects that have 3–4 years of timeline for execution suffer at least 1 or 2 year delay due to various reasons. In 2004, steel prices had almost doubled. Contractors who executed their projects during this time without an escalation clause in Contract would have incurred heavy losses, provided that the Project's Material was mainly steel oriented.

If the escalation clause and formula are provided in the Contract, it could minimize unnecessary arguments/disputes. Owners seek to have the lump sum component of the Contract Price remain unchanged from the commencement to the end. Otherwise, the Owner will not be able to meet the Investor's projected Internal Rate of Return (IRR). Often it is difficult for the Owner to accept an increase in cost from the Contract Price, which cannot be absorbed by contingencies built into the Contract Price. When the loss-making scenario confronts the Contractor, his cash flows become possibly negative, the Contractor will not be able to procure consumables and materials on time, and he may cut down overtime to workers. If the Project experiences a delay caused due to a severe loss-making situation to the Contractor, the bigger loss will confront the Owner due to loss of revenue for the late start of commercial operation. Such losses to the Owner could increase IDC (interest during the construction phase) and reduce the NPV value or IRR of the Project.

From the IMF, under the heading of "Inflation, consumer prices (annual %)," the annual shifts in the consumer price index in percentage are shown in a graph. In 2016, a particular country XYZ recorded 8.06%, and this 8.06% is an increase from the scheduled completion time of the year 2015 for a Project in that country. Then the Contractor should assess the value of the remaining portion of work from the start of the year 2016 under each WBS item/sub-item from the Progress Measurement System.

Let us say the value of remaining work after 2015 was 100 MUSD, which was subjected to inflationary pressure, then $100 \times 8.06\% = 8.6$ MUSD could be roughly the sum to be considered as the escalation cost payable to the Contractor if the delay reasons are attributable to the Owner alone.

However, in most cases, both Parties contribute delays; hence the win-win formula could be 50:50, which makes 4.3 MUSD the justified amount. The Contractor has to identify the cost items that are directly affected by this, that is to say, if the engineering of this project is executed in a country away from the country XYZ, such part of remaining work after 2015 cannot be factored/accounted for the calculation for this claim.

If guided by a professional organization, the accuracy of such calculation can be improved.

23 Liens

The origin of the word 'Liens' can be traced to a Mechanic's Lien. In a construction contract, the word Mechanic's Lien seems odd. In those days of the 16th century, anyone, whether a welder, carpenter, plumber, or general labor, who did physical labor has been called a Mechanic, and this term continues even though the construction industry has made significant technological advancement over generations. A lending bank can enforce a lien on a house property (offered as a collateral asset) belonging to the loan borrower as a guarantee for future loan repayments, failing which the bank can exercise its legal right under lien over the personal property.

Owner, in Contract, includes clauses to prevent any claim on its personal property, which may comprise the Plant, Facilities, Site, or property on which a Contractor has performed work during the Project period through the engagement of Subcontractors by Contractor, as a lien is basically claimed on the personal property.

In a contract execution, the question of lien arises only at final payment release when the Owner requires a Claims and Liens Release Letter. The final payment represents the final contract price with adjustments made for an agreed extra price on approved Change Orders, approved claims, back-charges/deductions, and any and all claims arising from a lien. If this Release Letter is not submitted, the Contractor can again possibly come up with a new claim, may be contractual or otherwise. A similar claims and liens release letter also applies to Subcontractors and Vendors of the Contractor.

A Claims and Liens Release Letter should be issued by the Managing Director or CEO or the authority that has the power to do so through notarized Power of Attorney.

A sample form of the Claims and Liens Release letter is shown in Appendix 8.

24 Total Liability

A difficult situation may arise for the Contractor during the negotiation with Vendors/Suppliers of the well-established category before awarding a Purchase Order to them. Such situations arise when the legal department of these Vendors raises deviations to total liability clauses appearing in the General Terms and Conditions (GTC) of the Purchase Order. Often the deviations surface due to a lack of clarity on the limit of liability under a worst-case scenario. No Vendor can or will assume unlimited liability. Let us list down the various scenarios wherein the Vendor/Supplier may become limitedly or unlimitedly liable. These liabilities are categorized into Category A, Category B, and Category C.

Category A

No limit in liability

Under this category, no limit can be stipulated for liability in case of violation by the Vendor. Limiting the liability is not possible for the following cases:

1. Indemnities to be provided to protect/indemnify a Purchaser (Contractor) arising from a claim from third parties to the Purchaser. Claim here means any claim arising from Suppliers causing bodily injury or death or damage to property of the third party. Insurances are procured to provide the protection/indemnifications.
2. Breach of civil code or law of a country involved or international laws and regulations.
3. Infringement of Knowhow (License, patent, intellectual property).
4. Breaching Confidentiality Agreement or NDA.
5. Code of ethics if incorporated in the Purchase Order conditions (if applicable) is violated.
6. Violating Sanctions imposed on some particular countries (if applicable).

Category B

0% liability

Under this category, there is zero liability on Supplier. A Purchaser cannot hold a Supplier responsible for any of the following, if the Purchaser is to face consequential losses arising from the actions of the Supplier in respect of

1. The financial or economic loss incurred
2. Loss of profit

3. Loss of production of plant arising from faulty or breakdown of Vendor equipment
4. Loss of contracts that the Purchaser may face, for which reasons can be attributed to the Supplier
5. Any indirect damages

Category C

150% liability

This is related to damage to Purchaser's property. Indemnity can be provided to protect/indemnify the Purchaser from losses caused by the Supplier causing damage to the Purchaser's property. Under this category, a Purchaser can hold a Supplier with some limit of liability, say 150% or alike. An Owner who is the end-user also seeks indemnity even if these are caused due to reasons attributable to a Supplier, where the contractual relationship is between a Purchaser and a Contractor.

25 Payment Terms

25.1 PAYMENT TERMS WITH OWNER

Owner makes payments as per the invoicing procedure to Contractor against the approved invoices. The Owner releases the first payment when the Contractor submits the bank guarantee bond.

The rest of the payments is paid monthly against the approved monthly progress statement, which has two components: The first component is for approving a certain allotted percentage for achieving a milestone (say payment milestone) and the second is for the physical progress measured based on the approved WBS (work breakdown structure). The Contract contains a table reflecting various payment milestones to be achieved by the Contractor each month. The payment milestones correspond to important/key activities of the Project. The Contractor also submits a Performance bank guarantee equal to 10% of Contract Price before seeking any progress payments.

If the first component were 25%, the second component would be 75%. The progress payment certificate is made after verifying the progress of each discipline every month in accordance with the WBS. Thereafter, invoicing is prepared based on the progress certificate signed and approved by the Owner.

25.2 PAYMENT TERMS WITH VENDORS/SUPPLIERS

The payment terms and conditions should be such as to enable Vendors to sufficiently meet their cash outflow. When Vendors face a cash crunch, they delay payment of salaries to employees and payment to sub-suppliers. Often such situations cause a strike or slowdown by workers and delayed delivery of consumables (welding electrodes and gases for cutting, etc.). On the other hand, more payment made to the Vendor than required could put the Contractor into a disadvantage position when a Purchaser intends to terminate the PO after cash-generating items have been delivered.

A sample payment break up is shown below (Table 25.1).

To generate cash flow at the beginning, an advance payment of 10% can be made upon submission of an advance bank guarantee or bond to the Owner, and this advance payment can be progressively deducted by the Owner at each progress payment. The bond value should be accordingly adjusted to reduce the fees that the Contractor is liable to pay to the bank. Payment for delivery also can be made upon submission of the material bond (bank guarantee) to the Owner and this bond can be released by the Owner when equipment and material consignment comes under the physical possession of Owner.

TABLE 25.1
Payment breakup for Purchase Order (PO)

Percentage	Payment Condition
10%	PO is accepted by the Vendor in writing. The manufacturing schedule is submitted and reviewed with comments. Key Vendor documents are received and commented by the Purchaser.
20%	Copy of Unpriced sub-orders is received for all major components (say forgings and plates). Sub-order is one that a Vendor places on its supplier or sub-vendor.
20%	Receipt of all major sub-ordered materials (plates and forgings) at manufacturer's shop (certified by third party engineer or shop expeditor). Spare Parts list with the unit price received and approved by the Purchaser.
40%	Upon complete delivery of equipment and materials (defined per INCOTERM 2010). Normally FCA or FOB, all depends on how a Purchaser organizes this and Insurances. The definitions of FCA and FOB are given in another section.
10%	Upon acceptance by Purchaser and PO close out completed.
100%	Total

26 Planning

For planning and scheduling, the main tool used is Prima Vera. Over the years, its version moved up, and the current version known is P6. The contract provides a dedicated section and basis for Planning and Scheduling. Contractor after award prepares a Procedure based on the Contract requirement. The Contractor creates templates to prepare the Monthly Status Report and Monthly Schedule Report in consultation with the Owner. In every monthly update done as per approved cut-off date, the Schedule indicates the remaining duration, start and finish date as per Baseline, and start and finish date as per Forecast for each activity. A Critical Path Schedule is separately shown with analysis on critical activities from a planned one to various updates. S-Curves, Tables, and bar charts made provide a proper illustration. Histograms on Manpower are made. Since Engineering progress is measured in man-hours consumed for each deliverable involved, a comprehensive table is made showing estimated man-hours for each deliverable in its various stages of progress. Cash flow curves are prepared but they constitute a confidential document and hence its distribution is strictly restricted. Work Breakdown Structure is an important task and this is prepared for each discipline. A dedicated table containing all the milestones is made and updated monthly with remarks.

A Procedure for Progress Measuring System (PMS) is made. A dedicated PMS for the following sections are made:

- PMS for Management;
- PMS for Engineering;
- PMS for Procurement;
- PMS for Transportation and Logistics;
- PMS for Construction;
- PMS for Pre-Commissioning;
- PMS for Commissioning, Start-Up, and Performance Test;
- PMS for Temporary Facility.

Chapter 27 supplements the above.

In the Contract, normally a Level 1 or 2 Schedule (Baseline) is shown under the condition that this schedule will be developed into Level 3 after awarding of Contract. Moreover, this Level 3 is called a Baseline Schedule (Detailed). Level 1 and 2 schedules are generally annexures to the Contract and are recognized as the Contract Baseline schedule or Work Time Schedule, and these schedules serve as the basis for the development of a Detailed Baseline Schedule after the commencement of Work. This being a Contract schedule, it is important to reflect the key activities, long lead equipment deliveries, and its field-associated piping construction work completion dates in the critical path.

In Level 1, one or two activities per Engineering, Procurement, Construction, Pre-Commissioning, Commissioning, Start-up, and Performance Testing Section are shown. Besides, Key Dates including Contract Deadlines are also shown.

In Level 2, the section mentioned in Level 1 is expanded to include sub-sections (or disciplines) such as Civil, Building and Architecture, Structural, Piping, Pipeline, Electrical, HVAC, Telecommunications, and Instrumentation discipline. Hence, the number of activities can move upward. Besides, Milestones are shown.

In Level 3, the sub-section (discipline) mentioned in Level 2 is expanded to include sub-sub-sections. Level 3 Schedule is explained in Chapter 1.

In Level 4, quantities are reflected under each discipline. Level 4 in Prima Vera Planning is normally performed by Sub-Contractors. However, the Contractor also prepares a Level 4 Schedule to reflect a 6-week-forecast when the Construction work phase reaches peak load. It assumes importance as the Owner can have clear visibility on the trend of the progress of activities. The visibility means, for example, the quantity of concrete to be poured, tons of steel to be erected, and length of electrical cables pulled at planned locations and areas. At peak load in Construction, this schedule assumes the highest importance and serves the best for the members of the project management teams of both the Contractor and Owner. Slippages in progress noticed in Level-3 can be very clearly and accurately mapped on Level-4 to increase the target of quantities for the subsequent weeks.

27 Progress Measurement

Progress quantity achieved per discipline against the planned quantity is measured as per the approved Procedure called the Progress Measurement System (PMS), especially the WBS. It is the Contractor's responsibility to have this PMS in place. PMS is made for Engineering, Procurement and Construction, and Commissioning as explained under the Cost Control section. Chapter 26 supplements what is stated in this chapter. Progress is reported weekly and monthly officially to the Owner.

27.1 ENGINEERING

In the case of engineering, each discipline under Engineering has a certain number of engineering deliverables to be made. In a Master Document Register, all these deliverables are populated for various stages: (1) issue for design (IFD), (2) issue for construction (IFC), and so on. For each issue, the activity should show certain man-hours. If the deliverables are approved, then the corresponding man-hours are counted to bring about a total. A percentage is obtained on the weight factor assigned under the WBS chart. The planned/progress quantity is measured in man-hour. Each engineering deliverable is allocated with some definite man-hours in the plan.

27.2 PROCUREMENT

Here also the progress is measured against a certain number of activities/deliverables. Unlike engineering, here each deliverable progress is based on the weight factor. For example, Receiving Quotation, Commercial Bid Evaluation, Vendor Print, and Fabrication (Manufacturing) are activities/deliverables. Delivery activity is treated as a separate activity under Logistics Discipline, although it is closely related to Procurement.

27.3 CONSTRUCTION

For construction, the measurement goes as per quantities, for example, steel structure erection in tons, foundation work by the cubic meter of concrete, and so on; such quantities are weighed against an allocated weight factor. Quantities can be estimated at the Level 4 Schedule.

28 EOT Claim

Let us assume that a serious delay occurs to the Project due to reasons attributable to the Owner. The Contractor finds that reasons are related to the Owner's review delays on drawing/document, late decisions, and reversal of decisions, interference, disruption, and lack of co-operation.

The Contractor who is affected thereof seeks to prepare an EOT claim dossier through professional consultants with time impact analysis done through Prima Vera scheduling software P6 for the losses he has faced in Engineering execution at headquarters and Construction execution at the Site.

In the execution of Engineering, the Contractor sees that he can make a claim to seek compensation for prolongation of engineering resources and inefficiencies cost he suffered due to Owner's review delays of engineering deliverables. Chapter 32 addresses review delays in detail.

In the execution of Construction, the productivity reduces, which consequently results in an increase in the man-hours. The Level 3 Detailed Baseline Schedule also reflects the total manpower (direct and indirect) planned for the Project, which can be compared with the actual man-hours consumed and reported to the Owner in Daily Report, Weekly Report, and Monthly Report. In the Gas Processing Industry, it is known that for a Project of 1 billion USD magnitude, the estimated figure could be in the range of 8–10 million man-hours. The Contract of Lump Sum Turnkey EPC format does not allow compensation for man-power increase from original estimates on which man-power histograms were submitted at the final phase of bidding. As such, the Contractor knows that he cannot contractually qualify the claim on a man-hour increase. Yet he can draw a sympathetic consideration from the Owner if he explains the increase of man-hour in a professional manner. Such strategies help at the end of Project completion when outstanding commercial issues come up for amicable settlement.

Contractor finds that the Site conditions as they appear in the following have changed from the bidding basis and hence sees the possibilities of the claim.

1. The Owner has delayed the access road; the Contractor's transportation task becomes difficult and inefficient,
2. Subsoil condition is harder to excavate, this means the soil data provided in the bidding stage is different,
3. Security issues in the country reduce the productivity severer than what was at the time of bidding,
4. Working hours restriction at Site, and
5. Lack of inspectors from Owner delaying the inspections.

In the history of Project execution, most of the Projects are seen to have suffered delays. Hence, right from the beginning of the Project, Contractor draws up claim strategies at the top management level. If the Contractor has nothing to justify his claim, then there is no case for a win even when the Owner is capable of understanding a possible overrun and loss-making situation to the Contractor.

The Project Manager continues to generate records for this purpose right from the beginning, even though he may have hopes that the Owner is unlikely to levy liquidated damages for delay. In several Projects, Project Managers lost their employment or were demoted at the end of the Project when Owners refused to settle the claims for lack of professional management of claims from the Contractor's end.

A scheduling strategy at the top management level assumes paramount importance here. To capture the delays for which reasons are not attributable to the Contractor, the Contractor shows in critical path whatever Owner's responsibilities as constraints in the schedule. The following can be designated as constraints in the critical path:

- Early access to Site by Owner (Constraint);
- Handover of all portions of Site (Constraint);
- Site preparations if it is Owner's activity (Constraint);
- Handover of Off-Site well pads or alike (Constraint);
- Feedstock availability (Constraint);
- Security (military and so on) support in countries where such support is essential throughout Project life;
- Electric feeders to provide construction power (Constraint) if this is Owner's scope, and;
- Water source for construction (use Constraint) if this is Owner's scope.

Extension of time is accompanied by cost compensation.

Prolongation Cost

Prolongation cost corresponds to the services rendered by indirect manpower, more precisely indirect cost and profits. This is also stated as overheads and profits. Overheads mean indirect manpower and other similar costs. These overheads have two components: one is head office overheads and the other is site office overheads. In some contracts, the overheads and profits are shown in the Price breakdown, which appears in annexures allocated for the Price Schedule in the contract documents. In case such breakdown is not shown, the Contractor should follow the International Doctrines to compute the overheads and profits. In Appendix 9 and Chapter 14, Prolongation Cost is explained in detail.

Disruption Cost

The disruption cost would mean any cost that may be caused as a result of disrupting an on-going work planned by Contractor arising from the actions of the Owner, and this cost corresponds to the direct expenses. In Appendix 9 and Chapter 14, Prolongation Cost is explained in detail.

29 Interpretation Rules

29.1 CONTRA PROFERENTEM RULE

The general principle in the court of law says Contractor's interpretation overrules Owner's.

The purpose of this rule is to provide justice to the Contractor as the Contractor is considered to be a weak entity in comparison to the Owner. Since Owner is the one who drafts or proffers the contract terms and conditions, the Contractor has little choice but accepts all such terms and conditions for the fear of disqualification in the bidding race among his competitors. Ambiguities in Contract may arise when a context addressing an obligation of Contractor presents two meanings regarding a certain work scope of Contractor, of which one meaning is advantageous to the Contractor and the other is advantageous to the Owner. The Contractor tries to interpret in such a manner that the scope of work in question is not his, while Owner says it is his. Each Party consequently engages in writing a series of contractual letters to disqualify the contractual argument the other Party gives. More often, this situation begins to ruin the Project execution atmosphere and the unity of project teams.

In some contracts, Owners have obtained prior consent from the Contractor prospect during the final negotiation in bidding to stipulating in the contract that the Contra Proferentem Rule does not apply. In such cases, it amounts to unethical business practice.

29.2 EJUSDEM GENERIS

If the scope section of Contract stipulates, "Contractor shall furnish the building with items such as air-conditioners, refrigerator, sofa, and etc....", then the word "etc." appears to include under its ambit whatever items falling under the family of members as said above. The Owner's engineer may assert to include as many items as possible within "etc." to be the Contractor's scope of supply. In opposition, the Contractor's engineer may provide his counter interpretation to deny. If a difference in opinion arises between a Contractor and an Owner in the execution phase, it leads them into writing letters of unnecessary arguments. Does Contractor's interpretation prevail over the Owner's? The answer is both Yes and No. Contractor's interpretation prevails if the Contractor has not agreed to the exclusion of the Ejusdem Generis Rule in the Contract. Furthermore, both parties should note that "etc." is a general word. In the interpretation, a general word cannot dominate over a specific word when the whole context surrounding the scope of supply in question is considered. An interpretation made in letter and spirit holds good always, while an interpretation made without spirit creates chaos.

In view of the above, the word "etc." should be avoided while enumerating the scope of supplies in the Contract.

For more reading, visit the website: https://www.law.cornell.edu/wex/ejusdem_generis.

29.3 EXCULPATORY CLAUSES

Exculpatory clauses are those which are designed by Owner to give undue advantage to him in contract terms and conditions proffered by Owner whereupon the Contractor during bidding stage has little authority or strength to seek a fair phrasing of such clauses. The undue advantage means pre-emptive escape routes designed in Contract. The intention of the Owner hereof is to prevent him from falling into unforeseen liability or to invalidate a claim of Contractor before going into the merits thereof. Such escape routes can be made by skillfully phrasing contract clauses. Clauses so designed become unfair and frustrate the Contractor whenever it prevents the Contractor from qualifying his genuine claims. Frustration breeds strife and corrupts the congenial relationships in the project execution phase. For further reading, the article from the following domain would help: https://contracts.uslegal.com/breach-and-remedies/exculpatory-clause/

30 Change/Variations/ Trend Notices

30.1 CHANGE

A Change is defined to cover any and all addition, deletion, and modification to the scope of Work.

A Change is one to which Parties agree in principle to reach settlement in cost and time impacts, whereas a Claim Parties disagree. Contractor understands that an approved Change can be settled within the time frame stipulated in the Contract, while an unapproved one cannot. Claims are resolved through a Dispute Procedure, while Change through a Change Procedure. Changes are handled by middle management, while the Claims by top management. As such, Contractor escalates an unapproved Change into Claim through contractual notice. When the Contractor notifies a Claim, he soon makes efforts to establish its legitimacy by highlighting the relevant clauses and giving his arguments. In the notice, he reserves his right to entitlement on cost and time in the future. He marks such letters with the heading "Without Prejudice".

When Contractor is willing to proceed with any work under Change, while the Owner maintains silence on approval that Contractor is expecting, proceeding with Work before approval of Change increases the risk to Contractor. Hence, the Contractor endeavors to reach an agreement before proceeding with the work covered thereunder. In industry practice, the Contractor is not expected to proceed with any Change unless approved in writing by the Owner.

If the Owner rejects in writing any Change requested by the Contractor but directs Contractor to proceed with execution of the works involved under the Change in question to serve his own priorities, a complex situation unfolds in the project execution. Though such a direction displeases and is unacceptable to the Contractor, he may not be in a position to disobey contractually this direction but he can improve his contractual position by asserting his right to entitlement while complying with Owner's instruction. The Contractor should immediately present an approximate cost and time impact. The Contractor should not fail to include this issue in the forthcoming Steering Committee meeting, which is attended by top management.

When Owner recognizes an item/work covered under a Change request from Contractor to be a genuine Change and in the critical path but is unable to finalize the cost and time impact within a reasonable time, then Contractor runs into the risk of executing it. If Contractor allows this status quo of Owner to continue until when the entire Work of the Project comes to completion, then Contractor will be in a disadvantageous position to negotiate the settlement, while Owner moves into an advantageous position. If Contractor stops work pending approval from Owner fearing disadvantageous position in the negotiation, Owner can hold Contractor for breach of Contract and liability to pay for significant damages. When the approval process of cost and time takes longer time for any work under the Change in critical path, the Contractor is not expected to halt the work. If the Contractor halts such work, it may erode the confidence and respectability, and such erosion cannot usher in a positive atmosphere when Claims come up for discussion and negotiated settlement. Often, cost approval takes a longer route and time, and hence rushing after settlement prior to starting a critical work item may not deliver the goods. If the activities under such circumstances are not in a critical path, it is prudent that both Parties complete the process of cost approval before commencing any work thereof.

In the fixed lump sum portion of Contract Price, the price is based on the given engineering data, soil data, and drawings and specifications. If the Owner reduces the overall Plant layout size during detailed engineering, such reduction could consequently reduce bulk materials such as cable lengths (electrical and instruments), piping lengths, and structural items (where pipe racks are also shortened). During the bidding stage, the bidder depends on this layout and prepares bulk materials MTO to determine the quantity and cost. In the bidding stage, pre-bid engineering is conducted through Engineering firms or in-house engineering departments who prepare MTO based on Bid Data. Another example is seen when the equipment size or numbers are altered by the Owner during DED. Since sizing of the equipment and piping are already frozen at the Front End Engineering (sometimes this is called Design Dossiers), any alternation entails a Change Order preceded by the engineering deviation approval.

The Owner is normally not responsible for an additional cost if the Contractor has made mistakes in the preparation of its material take off for the purpose of bidding. The Owner does not require a Bill of Quantity information from the Contractor at the bidding stage for Lump Sum EPC format of contract. However, the Contractor should prepare to accurately perform the cost estimate.

Where time impact is associated with cost impact, it is advisable to delink the time impact from the Change Request. By delinking, the Owner is freed from administrative hassles to approve the Change. Once approved, the Contractor can issue PO to Vendor for procurement and supply of the critical equipment and materials. Time impacts so delinked are gathered over a period of time. When the appropriate time comes, the Contractor can take up the case. In industry practice, approving time extension during execution is not an easy task. Due to the complex nature of Project activities, any additional time approved at the early or middle stage of the Project can be found inappropriate at the last stage. For this reason, the Owner delays his approval on time extension until the last stage of the Project. If larger delays have occurred, then it is prudent for both Parties to review and readjust the Contract Deadlines mid-way including the applicability of Liquidated damages. Readjustment of Contract Deadlines consequently requires application of modification to the existing Level-3 Baseline Schedule and Key Milestones and Payment Milestones tables.

30.2 VARIATION

The word "Variation" refers to the "Change". Under FIDIC Conditions of Contract for Construction, applicable for building and engineering works, the word Variation is used to mean Change. To remove difficulties in the contract administration of Projects, which is based on Lump Sum Turnkey EPC Format, it is advisable to avoid multiple terminologies while drafting contract terms and conditions. The term Variation can be dispensed, provided that Change clause thereof is defined appropriately.

30.3 TREND NOTICES

If the Contractor's anticipated profit is eaten by higher procurement cost of equipment and materials or labor unrest during Construction, the Contractor has a tendency to explore options to generate a list of Potential Changes and converts them into Changes when he finds appropriate time. Owing to the smaller magnitude of the cost involved, the Contractor has a lesser concern when any Potential Change surfaces in the Project, but allows compilation of all Potential Changes until the Completion of the Project. The summation of all Potential Changes sometimes certainly gives a bigger magnitude of cost, and the Owner has difficulty rejecting it if the Contractor is able to justify it. The Owner also develops a similar list but of negative Potential Changes to offset the Contractor's sum. To professionally manage the above situation, a Trend Notice Mechanism for Potential Changes is advised as below.

If a Project is hugely financed through lenders, then lenders expect the Contractor to keep a track record of Changes that may impact the cost and timeline of the Project through what is called the Trend Notices Tracking Mechanism. Such a mechanism includes all kinds of Changes, Potential Changes, and Variations in order to have all under one umbrella.

As it is important for the Owner to have a notion of what the Contractor has in mind with respect to Potential Changes, the Owner should encourage the Contractor to even notify any Potential Changes by issuing Trend Notices, which are different from Change Notices. Owner and Lender can have visibility on the trend captured under Trend Notices.

31 Contract Price below Floor Price

At the time of final negotiations before the award of a Contract, some EPC bidders try to reduce the price below the floor price in order to secure the win. For an Owner, a reduction in EPC Cost certainly reduces the Project Cost, which in turn increases the financials measured through IRR, NPV, and DSCR.

If the EPC Contract Price at the award is found to be 10% below the floor price, the Contractor may try to transmit this loss to his Vendors by reducing PO price in the procurement of equipment and material. Such reduction may be possible only if the Contractor awards POs to less reputed or financially wreck Vendors who are looking for survival in the distressed situation. PO awards to such Vendors not only result in delays in deliveries but also increase cost eventually as explained below.

When the deliveries of the Vendor in question fall into criticality and the overall Project completion date shows delays consequently, the Contractor comes under pressure. The Contractor succumbs to the pressure and agrees to the demanded cost of the Vendor in the backdrop of upcoming Pre-Commissioning activities in the critical path and pressure from the Owner on the delayed delivery. At this juncture, he is rendered toothless in the cost negotiation with his Vendors when they demand approval of cost claimed in Change Request. The increased cost on PO occurs at this time and is inevitable. Finally, the Vendor can reverse the loss loaded into the original PO Price.

32 Owner's Delay on Document Approval

In the history of delayed Projects, it is possible to find Owner actions delayed on several fronts. The following is one among the many.

If the Owner were to show an inconsiderate attitude lacking the spirit of co-operation, the Contractor would find the approval of Documents taking more time than it should. If the Owner were to find a new comment and demand further revision at every submission, the cycle might run endlessly. Such a situation is not only painful to the Contractor but also harmful to the Project.

The remedy can be found under the following conditions:

- The Owner takes no more than 7 working days to comment in the first round of Documents submitted by the Contractor;
- The Owner refrains from raising new comments in the second round of submission;
- Both Parties agree mutually on new comments raised by the Owner in the second-round submission, provided Parties understand that such an instance is special and important;
- At the end of the review by the Owner on the second-round submission, both Parties should meet and reach a consensus for closing all outstanding comments. The conclusions can be wrapped up in the minutes of meeting.

It is necessary to classify Documents into three or four categories as explained in Appendix 7. Category 1 means Contractor cannot proceed with any work related to the Document on which Owner does not give approval. If Contractor disregards this rule and executes work, it results in non-conformity and Quality discipline issues NCR thereof.

In a Contract based on the Lump Sum Turnkey EPC format, even if Owner approves a Document, which is later found to be defective and such defective Document has become the cause of some accident and damage to the Work executed, Contractor cannot escape from his responsibility to remedy the situation at his own cost and risk. No approval from the Owner can excuse the neglect in the Contractor's responsibility to provide a defect-free Document. Being so, the Owner, without compromising the quality, should moderate his approach in the review and approval of Document, unless otherwise such comments assume paramount importance. In other words, a balanced approach between quality and time is recommended.

33 Insurances

The Insurances applicable to the Project are indemnity based. It covers both fundamental risks and particular risks.

Contract Manager in the Project Management Team becomes the focal person to address the issues related to insurances during execution. However, an Insurance Specialist in a Corporate office is the proper personnel to address this issue and give final decisions when confusions arise or critical situations so warrant. An Insurance Specialist is not part of the Project Management Team, and he supports all Projects from the Corporate Cell.

For the procurement of insurance, territory definition assumes importance and hence a definition is provided below.

"Territory" is a country in which the Site of the Project is situated. For example, the Kingdom of Bahrain can be a Territory if the Site of the Project is situated there. Territory normally means to include land and water or sea boundary limits.

Generally the Owner, Contractor and his Subcontractor, supplier, Vendor, licensor, consultants, employees, and alike are "Insured".

For the procurement of insurances, there are professional consultants in the market and they know what best terms and conditions are available to optimize the premium. A Corporate Insurance Specialist helps the Project Team in procuring the insurances. Also, the Project Management Team is aware of the terms and conditions and the main features of Insurance policies. Insurance claims fail to qualify if the terms and conditions are not followed; for example, if the Contractor does not unpack the equipment within a specified time frame after its arrival at Site to assess the damage during transportation and report to the insurer, any claim raised belatedly on damage occurred during transit can be rejected by the insurer. The Project Management Team is responsible for the extension of the coverage period and amendments if any required in the policy.

33.1 EAR (ERECTION ALL RISKS)

The EAR insurance covers the full replacement value in the event any destruction takes place, provided this destruction takes place within the Territory. Full replacement value means the Contract Price. However, if the cargo during transit in the sea is damaged, then the insurance cover is covered under Marine Cargo Insurance, not under the EAR.

For some period following the Commencement Date/Zero Date, the Site generally has very negligible activities, often called Early Works. The Early Works may include soil investigation, site preparation, and construction of temporary facilities. The scale of risks potentially associated with the execution of Early Work is far less than those associated with the execution of Main Work. For a lesser risk, operating a bigger insurance like EAR is not an economical approach. Therefore, to reduce the premium fee payable to Insurer under the EAR, the EAR period can be shortened by the period during which Early Works takes place. In view of this cost-saving action, an alternate EAR policy is used for the period during which Early Works are executed. The alternate EAR period can also be stretched until the first lot of permanent equipment and materials start arriving at Site. In this cost-saving approach, the Contractor has the opportunity to optimize the premium fee payable to the Insurer.

Premium fee payable to an Insurance Company becomes less when the deductible amount is increased. Since this deductible amount of a claim by Contractor to Insurer cannot be paid by the Insurer, the Contractor, as a recourse, covers any risk falling under this deductible range through Subcontractor's scope of insurance. For this purpose, the general terms and conditions (GTC) of Subcontract addresses this scope of insurance on Subcontractor. The deductible amount is reflected in the insurance policy.

33.2 THIRD PARTY LIABILITY INSURANCE

By this policy, the claims on compensation or damages arising out of death, injury, or property to a third party within the Territory can be obtained from Insurers (Insurance Company). Normally, the first party is the Owner, and the second party includes the Contractor, his Subcontractors, and personnel, Vendors/suppliers/consultants/architects.

The following example is cited to identify who is the third party among others when a claim situation arises.

If a person's property in the neighborhood village is damaged by the movement of a concrete mixer or heavy trucks belonging to the Contractor or Owner, where this affected person, for example, is a local resident in the village and not an employee of Subcontractor/supplier/Contractor/Owner, then this person is called a third party for whom the compensation will be paid under this policy.

33.3 MARINE CARGO INSURANCE

This is to cover the risk of imported cargoes (say equipment and materials from Vendor shops) damaged during their transit from Vendor shops (Ex-works) to the Site/Laydown area. As per the current industry practice, the Contractor prefers FCA to FOB. However, such preference is left to the sole discretion of the Contractor in the Project execution. FCA is the abbreviation for Free-Carrier-(named place of delivery). For more explanation, readers may visit the web page: https://en.wikipedia.org/wiki/Incoterms.

Furthermore, marine insurance is required only at a time when the manufacturer starts sending deliveries related to ocean voyage. Considering the manufacturing time after PO issuance to Vendors, deliveries from Vendor shops do not happen for some considerable period from the Commencement Date of the project. Hence, the Marine Insurance cover is not required for the period preceding the despatch of deliveries. In view of the above, an interim insurance is a viable option to reduce the premium amount payable to Insurer.

33.4 VENDOR INSURANCES

Equipment and materials manufactured/fabricated in Vendor shops are separately insured by Vendors, as the EAR cannot cover risks at shops that are normally located away from the Territory covered by EAR.

Contractor obtains a certificate of insurance from the Vendor immediately after issuing the PO to Vendor. If the Vendor fails to provide the certificate, the Contractor can delay the payment to ensure compliance. By this certificate, indemnities are duly provided to the Owner and Contractor against any liabilities arising out of Contract or Subcontract (PO). The certificate of insurance proclaims Contactor as the certificate holder. Additionally, the certificate so obtained shows the Contractor as an Additional Insured to pursue a future claim. In the Insurance policy related thereof, an endorsement appears with subrogation waived. The purpose of this waiver of subrogation is to prevent the Insurer (of Vendor) from filing any claim from the Party who caused some negligence, and such party may sometimes include Contractor if he caused negligence.

33.5 OTHER INSURANCES

Other insurances cover the risks in respect of the following, which Contractor or Subcontractor owns/employs during construction:

- Construction equipment (cranes, excavators, DG sets, etc.);
- Automobiles (trucks, cars, jeeps, etc.);
- Any barges/vessels that operate on water and any aircrafts, and;
- Workers/employees.

Section III

Technical Aspects

In this section, there are ten chapters on the following topics:

Life cycle cost analysis (LCCA); owner delays on rigless Test; non-spec feedstock for commissioning; gas in and authority check points; value engineering; design defect and construction defect; NDT (non-destructive testing); destructive testing; hydro-test; heavy lift load sharing cranes.

34 Life Cycle Cost Analysis (LCCA)

LCCA is not intended for all equipment but for pumps and compressors. It is important for the Owner to understand the life cycle cost rather than the mere initial installation cost, the CAPEX. LCCA is not seen to be a matter of concern to the Contractor as this item corresponds to the operation period after the Contract has been executed and discharged. The procurement cost of a particular pump from a Vendor may be cheaper in comparison to another Vendor but it may prove costlier if the energy consumption, operation cost (labor), maintenance cost, downtime cost impacting on production, disposal cost (in consideration of environmental issues), and decommissioning cost are accounted for. However, these aspects are to be understood and analyzed while preparing the process data sheets for equipment and Vendor list in the Front End Engineering/Basic Engineering Stage. An EPC Contractor estimates his price during bidding based on the process data sheets, related Technical Specifications, and approved Vendor list. EPC bidding prices will go higher if the bidder gives attention to OPEX rather than CAPEX.

The website http://energy.gov/sites/prod/files/2014/05/f16/pumplcc_1001.pdf provides literature with the title Pump Life Cycle Costs: A Guide to LCC Analysis for Pumping Systems; article by S. Rahman and Dana Vanier (www.irbnet.de/daten/iconda/CIB9737.pdf) is also useful. Among others, Solomon Associates provides RAM Study reports to the FEED Contractor and Owner at the FEED stage to make decisions on the selection of equipment. The RAM Study is one of the engineering deliverables under Safety Engineering during the Detailed Engineering Design of Contractor in the Project Execution Stage.

For the implementation of LCCA during Detailed Engineering Design, the Owner makes sure that such requirement is stated in each equipment data sheet in the FEED stage and that Contractor during the Project Execution phase obtains Vendor quotes with and without LCCA costs before seeking Owner's approval of technical bid analysis of Vendors. The Contractor issues a Purchase Order to Vendor only after the conditions stated above are satisfied. Furthermore, the Contractor while making the Material Requisition provides a format on this so that the Vendor can comply with it.

35 Owner Delays on Rigless Test

In upstream gas projects, the wellhead profile after the rigless test can vary significantly from what has been provided in the FEED stage. The feed gas is a mixture of gas coming from a group of wells, which are 20 for the Project assumed. The gas as they come from wells enters flow lines, clusters, and trunk lines before arriving at the central processing facility. The feed gas quality depends on how good and bad wells behave. If the designed Plant does not have the flexibility to admit more gas from those wells that produce good quality gas and higher flow rates, then the final mixture entering the central processing facility may not meet the specification and this will be a bad news to the Owner. Consequently, the commissioned Project may not succeed fully in the commercial operation due to bottlenecks arising thereof. Some scenarios in this context are presented in the following:

- Surplus Gas: If some wells produce more gas than they should as per design, the pipelines designed without flexibilities cannot possibly handle it. To identify where the limitation in the designed pipelines exist, Contractor can have a flow assurance study performed through competent third party engineers with an objective to increase the flow rate in the designed pipeline and especially the choke valves installed at the wellhead. Often, this kind of remedial measures brings no workable solution if the pipelines have been installed.
- Underproduction: If wells produce lesser gas than what is envisaged in the design, this situation could make a commissioned Project commercially unviable, such as risks to Owner/Stakeholders of the Project. The underproduction of wells is linked to the condition of reservoir of gas.
- Quality of the Gas: If the gas from wells contains CO_2 content higher than envisaged in the design, the Acid Gas Removal unit cannot admit the full gas. As a result, the Plant throughput cannot be raised to the design level, and consequently, the revenue of the Project can fall even below the break-even point.

To avert the pitfalls that can kill the commissioned project and ruin the investment of stakeholders, drilling activity of wells should be scheduled to finish in all respects including rigless testing at least before the commencement of Detailed Engineering Design if not earlier. If the updated gas profile after the rigless test is made available to the Contractor before DED or at early stages of DED, then the Contractor can provide remedial measures through modification of the design of pipelines different from FEED and thus remove the bottlenecks resulting from the off-spec gas conditions. If the remedial measures require enhancing the capacity of the Acid Gas Removal Unit during the early stage of DED on account of the presence of CO_2, then simple modifications can be incorporated in the process without incurring higher CAPEX.

36 Non-spec Feedstock for Commissioning

It may so happen that the inlet gas or feed gas properties and characteristics may vary as said in the forgoing from design and upset those activities that fall under Contractor's scope of work related to Plant Start-up, load up, and normalization activities. In the Plant size we described in the beginning, the Contractor may find a timeline of 3 months between Gas In and Performance Test but the off-spec gas situation certainly makes this program impossible to complete within 3 months. If the Owner decides to have the non-spec gas admitted into the process, the Contractor should co-operate. When the Contractor decides to co-operate, he makes it clear in writing to the Owner that such co-operation does not prejudice his contractual rights and remedies. In the following, commercial impacts and technical solutions are investigated.

1. Commercial Impacts:

Erection All Risk Insurance Extension: If non-spec gas is admitted it could lead to some unexpected upsets and accidents. Then to replace new equipment, the Project may face delay as long as 1 year or even more. As the EAR insurance procured by Contractor continues to remain valid until a performance test is done, any replacement of equipment is through the coverage provided by the EAR insurance, but for maintaining the EAR longer period than contract base due to reasons attributable to Owner, the Contractor incurs not only the additional cost but faces a situation wherein his liability continues to remain tied indefinitely.

Performance Bank Guarantee Extension: Contractor extends the period of performance bond bank guarantee. The bank guarantee that the Contractor procured is for 10% of Contract Price. For 1 billion USD Contract Price that we assumed in the beginning, the amount stands at 100 million USD. The Contractor for the extension of this bank guarantee incurs additional cost.

The Indirect Cost: Contractor, in view of an extended period of stay, incurs an additional cost in this category to maintain his team at Site and a support team at HO.

2. Operating Plant with Off-spec gas and lesser loads:

When the Owner finds that the gas profile of the reservoir varies from FEED, he provides updates to the Contractor after undertaking adequate rigless testing. In FEED, estimated profiles are normally given including expected gas composition and pollutants.

If the feed gas composition, especially when the presence of CO_2 exceeds the anticipated value, differs from the design, AGRU cannot allow the maximum flow rate at the inlet due to its limited capacity to treat CO_2. The Plant consequently has to scale down its operating load. Under similar circumstances, one may find Plants operate as less as 60%–70% of full load design capacity.

For the Owner, such a scenario is the most troubling one as the Plant will not be able to generate profit. Under these circumstances, the Owner may try to have experimental runs to find out the real bottlenecks in the operation mode rather than through theoretical calculation. Owner at this juncture depends on Contractor's co-operation to operate the Plant with off-spec gas as much as the Plant is capable of accepting.

As the Owner is the one who invests in this Project, it is wise for the Contractor to understand the difficulties of Owner and support them instead of flexing muscle contractually but at the same time, Owner takes responsibility for any consequences that may arise out of using off-design gas for gas treatment.

The co-operation that Contractor extends helps create an amicable atmosphere. Contractor can capitalize on the situation that prevents him from conducting performance test runs and securing a completion certificate, for which reasons are attributable to the Owner. To reciprocate the Contractor's co-operation, the Owner's consideration to settle the Contractor's outstanding claims and issues will be available.

If the reader is interested in the process block flow diagram of AGRU (amine gas treating), he can find it on the website: https://en.wikipedia.org/wiki/Amine_gas_treating.

37 Gas In and Authority Check Points

Statutory authority checks the following after Contractor's submission of HSE Dossiers for Gas In and this can be called the PSSR, Pre-Start-up Safety Review session.

1. Off-Site well pads

   Safety systems are checked including Emergency shutdown valves, Safety relief valves, fire detection, and communication with Control Centre Room.

2. Main Plant

   The gas from the wells is allowed to move from the well towards the Main Plant. Before arriving at the Main Plant, it arrives at various sections such as trunk lines, slug catcher, LP separator, flare knock out drums, and fire water tanks. As the gas enters various sections, the testing is repeated for checking the integrity and functions of:

   a. Fire and Gas systems;
   b. Deluge systems;
   c. Firefighting with a hydrant hose;
   d. Fire detection and alarms, and;
   e. Public Address and General Alarm Systems.

38 Value Engineering

Contractor performs Value Engineering during or before the commencement of a detailed engineering design. It is the EPC Contractor who can undertake such engineering endeavors after the award of Contract. A few EPC contractors are known to have proven capabilities even to change the scheme of the process provided by Licensors: for example, a two-train-process had been reduced to a single-train saving enormous amount of cost incidental to compressors. Owners are known to have given credit to Contractors for the contribution of the latter in reducing CAPEX cost in the selection of equipment and material through his optimization of Project Specifications. Contractor gives importance to Value Engineering to further his profit, if the profit foreseen at the award of Contract is thin. The extent of credit that Contractor gets for such efforts is usually 50:50. In other words, if 4 million USD is saved, Contractor can get his share of 2 million USD. However, Contractor is cautioned not to undertake the Value Engineering task if Owner does not give away the credit. During the execution of Value Engineering, the Engineering department holds in abeyance certain activities of deliverables related thereof, which may consequently give rise to delays and impacts on the Schedule. In the bidding time, Contractor does not normally provide a window in the Level 3 Schedule for performing Value Engineering studies before releasing critical engineering deliverables. The activities of critical deliverables include the first issuance of P&ID, first Material take off on bulk materials (piping), Equipment Data Sheet, and Requisition for Long Lead Equipment.

39 Design Defect and Construction Defect

The design life of a building is normally 20–25 years; precisely the life is determined by the governing design code or Project Specification. In a Project, the design life of the building is counted from the moment the Final Acceptance is accomplished for the whole Plant. A Final Acceptance Certificate (FAC) follows the Provisional Acceptance Certificate (PAC). The warranty period is counted from PAC, and this period is normally 2 years or as specified in the Contract. Upon completion of the warranty period, the Owner approves the Final Acceptance Certificate. Figure 39.1 below reflects the following Milestones.

Milestones

- Zero Date: Start Date of the Project
- PAC
- FAC
- Design Life

Let us assume that a design defect leads to the collapse of a building due to faulty design from Contractor, and this collapse occurs after obtaining the Final Acceptance Certificate. The nature of damages falls under tort liability, not under contractual liability. Owner seeks remedy through arbitration, as the arbitration clause in Contract fortunately and expressly stipulates that arbitration shall also cover any claim for professional negligence or tort liability.

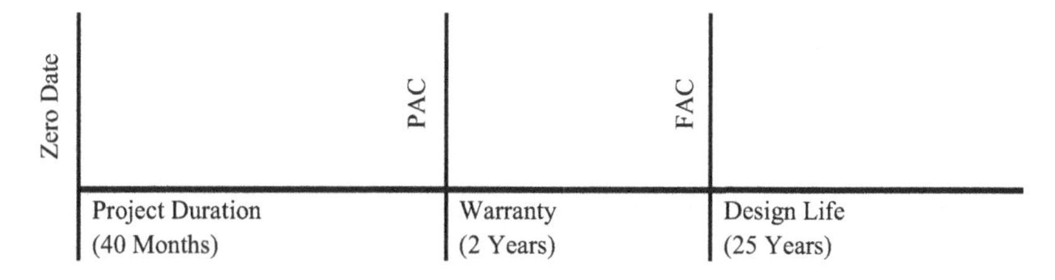

FIGURE 39.1 Timeline chart for Zero Date, PAC, FAC and Design Life.

40 NDT (Non-Destructive Testing)

NDT in oil and gas, petrochemical, or power plant project is generally used to identify defects caused during welding in metallic piping, tanks, spheres, and structures. However, in some cases, this method is also used to identify defects prior to using steel plates and sections in fabrication. To perform the NDT at the Site, professionals qualified in ASNT Level I and II are employed. They conduct and evaluate results. Level I personnel are permitted to conduct the NDT, while Level II personnel are permitted to interpret the results and provide an evaluation of defects. Level I and Level II personnel report to the QC Manager. In some Projects, the Owner requires the QC Manager to be Level III certified. Contract documents generally specify the extent of NDT requirements. For example, some weld joints of critical piping lines are to be 100% tested by gamma-ray. X-Ray is rarely used. X-Ray is limited to some special requirements. Accuracy and details of defects are more visible in X-ray-based radiography than in gamma-ray radiography, but X-ray is an expensive method. In the oil and gas industry, only gamma-ray radiography is used, while in the aerospace industry, X-ray may be widely used.

For carrying out NDT on some alloy or special steels or higher thickness carbon steel, PWHT (post-weld heat treatment) is necessary. The PWHT conditions are stated in WPS (Welding Procedure Specifications) and PQR (Procedure Qualification Records) related thereof.

The NDT tests generally used are as below:

- Ultrasonic Testing (U/T);
- Gamma-Ray radiography;
- X-Ray radiography;
- Liquid Penetrant test, and;
- Magnetic Particle test.

41 Destructive Testing

A tensile test for a weld specimen is carried out to qualify certain Welding Procedures, and it is done in the laboratory. The specimen to be tested is stretched by application of forces at either end. The specimen breaks off during the test after crossing the yield and plastic limit.

In the case of a concrete test, cube specimens of concrete are tested for measuring the compressive strength. Concrete cubes fail to give the desired results if the mix is not properly done. Since the cube tests are permitted after 28 days of curing, the Contractor should consider conducting the test at least 5 months before the scheduled date of the first concrete pour comes up. Three consecutive failures mean the Contractor loses 3 months in this testing activity.

42 Hydrostatic Leak Test

A hydrostatic leak test is a method of testing piping, pipelines, tanks, and vessels using water as the test fluid. At the Site, such a test is for piping/pipelines, and at Vendor shops, it is for vessels, where vessels include pressure vessels, heat exchangers, columns, and reactors. The word Hydrotest is often used to refer to a hydrostatic leak test.

Before conducting a Hydrotest of piping in the Site, the spool drawing or sketches are prepared by the Piping discipline (construction) and approved by the QC department of both the Contractor and Owner. Test-packs refer to the documentation, which includes all the technical information of the test spool and the NDT results with weld-maps. The test pressure is as per ASME B31.3 Chemical Plant and Petroleum Refinery Piping or Project Specification. However, each isometric or line list prepared by the Contractor indicates the pressure test value and the medium. A Hydrotest is performed in line with the Hydrotest Procedure approved by the Owner. At the start of the Project, a Hydrotest Procedure for above ground and underground piping and pipelines is made. Hydrotest requirements for pipelines are covered under ASME B 31.4 Pipeline Transportation Systems for Liquid Hydrocarbons and Other Liquids.

Floating roof-type tanks are fabricated at the Site and their purpose is to store water meant for process, utilities, and firewater applications. These tanks are tested for leak-proof with water filled in and held for a specified amount of time as stipulated in Inspection and Test Pan related thereof and the design drawings.

43 Heavy Lift Load Sharing for Cranes

When a heavy lift erection of equipment requires two cranes of lower and higher capacities in a tandem operation for sharing the load, there is a necessity to calculate using the moment's equation.

An example is given to illustrate a situation where a bustle pipe main of 150 tons is considered for a tandem operation in a blast furnace construction. The load for the lower capacity crane should not exceed 50 tons, while for higher, it is 100 tons. A Load sharing principle in the construction Site helps if the Contractor does not have a single crane of larger capacity. Tandem operations are a risky method in Construction, as the rigging operations of two cranes need to synchronize, for which a foreman stationed at a vantage point sends manual signals to the crane operators in their respective cabins.

Two cranes A and B are planned to erect a Bustle Pipe Main. Bustle Pipe Main is an annular duct made of heavy steel pipe lined with firebricks. Bustle Pipe Main, when viewed from the top, looks like a ring header made of pipe. It surrounds the Blast Furnace at a height where tuyeres are positioned to connect the Blast Furnace and Bustle Pipe Main.

Assume the Bustle Pipe Main weighs 150 tons and has a diameter of 40 m. Two cranes are available for performing the erection of the Bustle Pipe, which is at the assembly area/ground level. The load sharing ratio for Crane A and B is stipulated at 50 and 100 ton respectively. If Crane A is hooked at one extreme end of the Bustle Pipe Main, where should be the hook for Crane B positioned?

To resolve, let us assume the Bustle Pipe Main to be a horizontal beam of 40 m length with the uniformly distributed load.

The moment's equation at sling point A is as follows:

$100\ \text{ton} \times (40 - P)\ \text{m} = (150\ \text{ton} \times 40\ \text{m})/2$, where P is the distance to be determined for positioning B Crane hook from the end, which is diametrically opposite to Crane A hook point.

By resolving the above equation, the answer P arrived at is 10 m.

For Crane B, there are two hook points as the hook point moves away from the extreme end by 10 m on the circular Bustle Pipe to carry more loads, while Crane A will have a single sling (Figure 43.1).

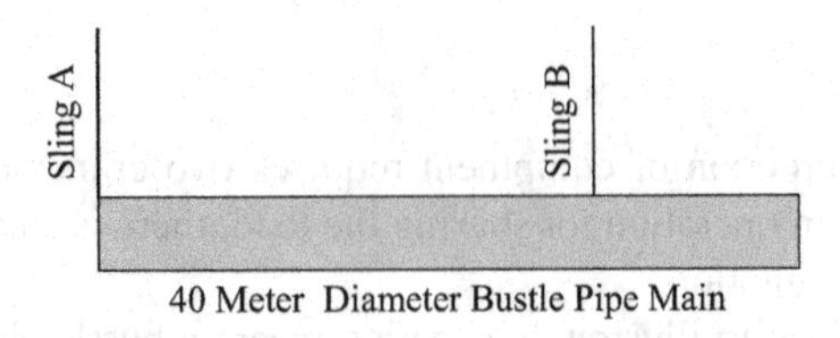

FIGURE 43.1 Diagram for Tandem Lift.

Section IV

Glossary of Abbreviations, Names, and Meaning

CISG Law: United Nations Convention on Contract for the International Sale of Goods.

Commencement Date of Project: this date refers to date on which the zero date commence and from which the project time line is counted. The date can be after the Contract is signed and made effective. In some projects, as agreed by the Contractor and Owner, the effective date of Contract and commencement date can be the same. However, commencement date cannot precede the effective date of Contract.

Defect Liability and Defect Liability Period: is the period commencing from the date of PAC until the date of FAC. The period as per Contract varies from 12 to 24 months depending on the Owner during the formation of Contract. During this defect liability period, Contractor appoints a Warranty Manager. He is responsible for taking action when defects are noticed. The care and custody of the Plant is transferred to Owner from Contractor at the date when PAC is recognized. The defect liability period is also called as Warranty Period. Owner returns the performance bank guarantee bond to the Contractor when PAC is achieved. At this juncture, Contractor submits another bank guarantee bond called warranty bond to Owner having validity until FAC is achieved. Warranty bond is generally half the value of performance bank guarantee bond. In general, performance bond is 10% of the Contract price, and warranty bond 5%.

Deliverable: means the output from executing a specified task; for example, in engineering, P&ID drawing is a deliverable, in procurement purchase requisition.

DSCR: debt service coverage ratio.

Effective Date of Contract: is the date on which parties enter into agreement and makes it effective or enforceable from this date onward.
Early Work Activities: are those activities which are required to be completed before the start of actual construction at Site. For example, building temporary office and labor camps, Site preparation work, boreholes, and geotechnical surveys are a few.
FAC: Final Acceptance Certificate is a very important certificate which Contractor obtains from Owner after successful completion of warranty period or defect liability period.
FEED: Front End Engineering Design. EPC Contractor normally relies upon this Front End Engineering for his development of this into DED (Detailed Engineering Design). FEED is provided by Owner to EPC Bidders during bidding stage. FEED is a design package specifying capacity of equipment, sizing of piping, the specification of feedstock, and products. FEED deliverables are mentioned in Appendix 5. FEED deliverables mean document and drawing corresponding to the FEED. FEED deliverables for licensed units are prepared by licensors, and non-licensed units are prepared by PMC (Project Management Consultants) appointed by Owner.
Feedstock: is the main raw material input into the Plant for production of products. Crude oil is feedstock for refinery; natural gas coming from wells are feedstock for gas treating central processing unit as explained in this book.
FIDIC: FIDIC is the International Federation of Consulting Engineers.
Guarantee: often confusion arises between the definition of warranty and guarantee. Warranty corresponds to workmanship of a product (equipment or field-constructed items), while guarantee corresponds to performance of the equipment or Plant in respect of intended output parameter for which the Plant is designed to operate. Warranty is limited to the defect liability period, which is 12 or 24 months after performance test runs are completed. Performance test is done to prove or verify that Plant is capable of producing the spec-gas and the spec-quantity and the limits on consumption of various utilities such as power, water, etc. For example, warranty for bearings replacement is specified for certain operation hours; if the wear and tear is quicker than designed, then Contractor or Vendor has obligation to provide replacements free of cost.
HP/LP flare: High-Pressure/Low-Pressure flare
ICSS: Integrated Control and Safety System
IFC: Issue for Design
IFD: Issue for Construction
IRR: Internal Rate of Return
ITP: Inspection and Test Plans
LCCA: Life Cycle Cost Analysis
LLI: Long Lead Items
LTIFR: Lost Time Injury Frequency Rate
MC: Mechanical Completion
MR: Material Requisition
NCR: Non-Conformity Report
NDA: Non-Disclosure Agreement
NDT: Non-Destructive Testing

Notice to Proceed: Notice to Proceed is the instruction or directive issued by Owner to Contractor to proceed with a certain portion or the whole of the work.
NPV: Net Present Value
Owner: is the entity who owns the project and takes over the Plant after construction and commissioning is done by Contractor.
PAC: Performance Acceptance Certificate
P&ID: Piping and Instrument Diagram
PMC: Project Management Consultant, normally appointed by Owner
PMS: Progress Measurement System
P.O: Purchase Order
PSSR: Pre-Start-up Safety Review
QA/QC: Quality Assurance/Quality Control
RAM Study: Reliability, Availability, Maintainability
RFSU: Ready for Start-up
TRIR: Total Recordable Incident Report
Subcontractor refers to the party who receives a Subcontract order for execution of work items within the premises of Site.
TBE: Technical Bid Evaluation
UNCITRAL: United Nations Commission on International Trade Law https://uncitral.un.org/en/about
Vendor refers to the party who receives purchase order from Contractor for supply of materials and services related to the project.
Warranty: see Guarantee
WBS: Work Breakdown Structure

Appendix 1
Project Organization

See Figures A1.1–A1.4.

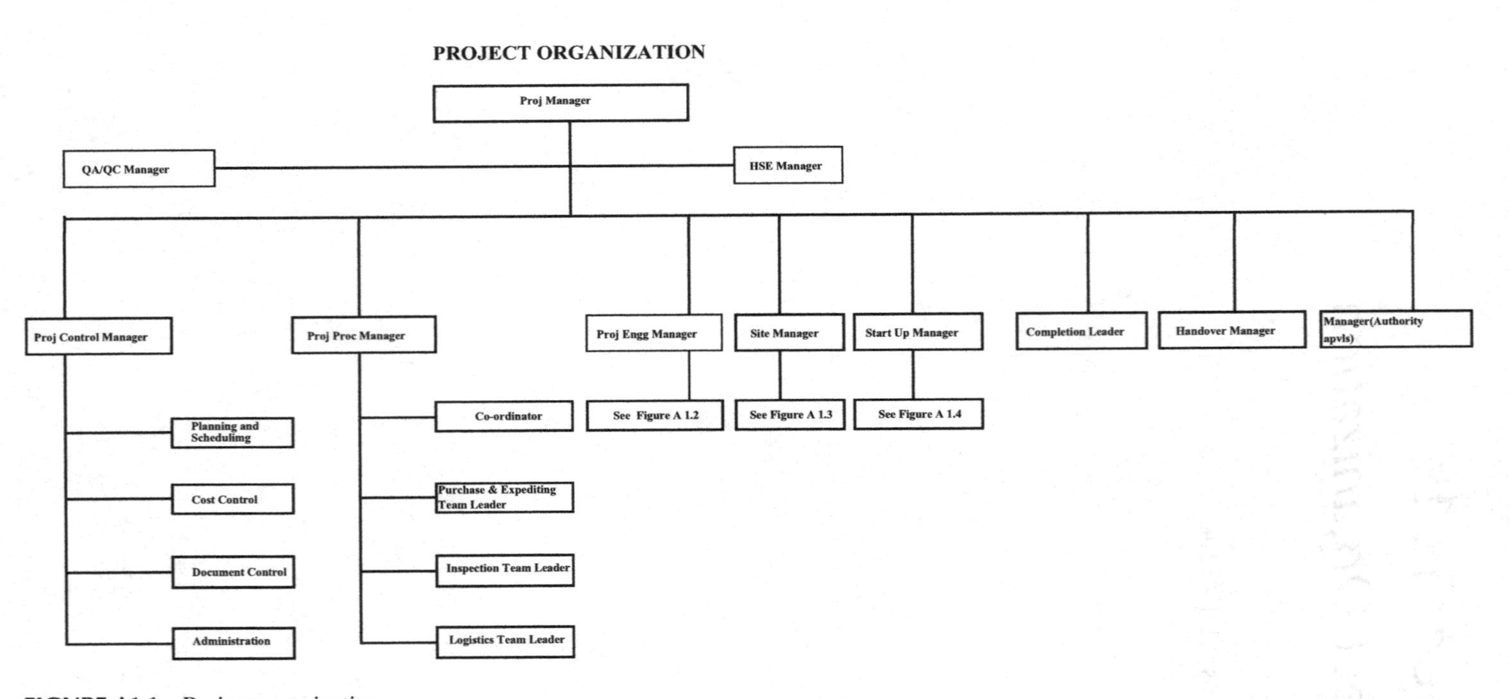

FIGURE A1.1 Project organization.

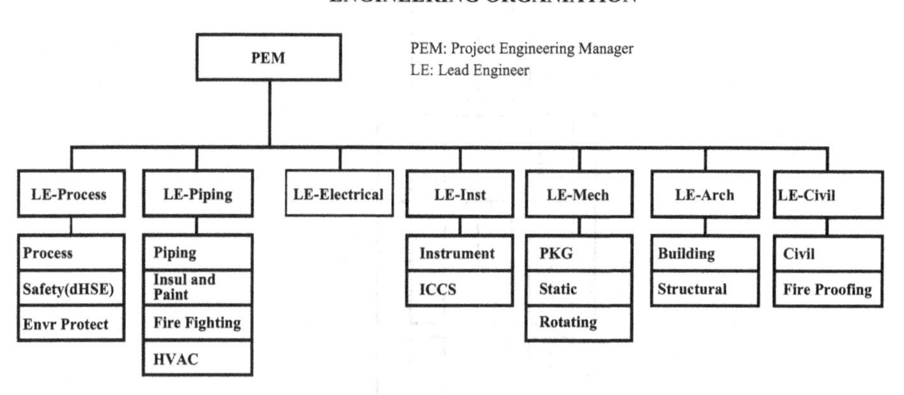

FIGURE A1.2 Engineering organization.

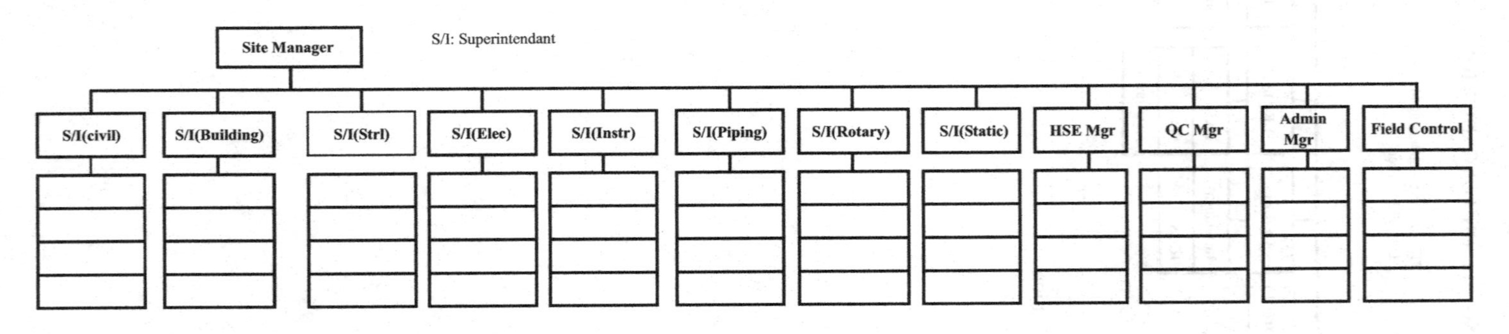

FIGURE A1.3 Site organization.

[a] The number of field supervisors or engineers under each S/I depends on the volume of work in each site.

[b] If Mechanical work volume is less, the one S/I would be sufficient instead of Static and Rotary S/I.

[c] Pipeline S/I can be added if the pipeline work exists.

[d] Offsite organization is not shown here due to limitations.

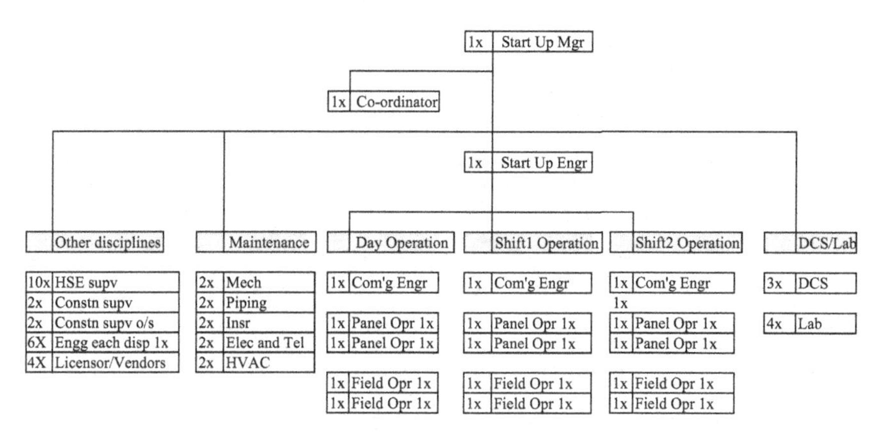

FIGURE A1.4 Start up and operation organization.

Start-up Manager is also the Commissioning Manager. Here commissioning means full Plant, while individual systems commissioning is the predecessor to Start-up. At Start-up, it is assumed that all utility and firefighting systems have been already put into normal operation and all individual systems have been commissioned after pre-com completion.

Appendix 2
Contract Price Weight Factor

See Tables A2.1–A2.3.

TABLE A2.1
Overall Contract Price Weight Factor

Cost Item (CI)	Cost Item Description	Weight Factor (WF)[a] (%)
CI 1	Management	3.27
CI 2	HSE, Social and Security	0.92
CI 3	Services/Assistance to Owner	0.20
CI 4	Engineering	7.57
CI 5	Procurement and Supply FOB	45.42
CI 6	Transportation FOB to Site	3.90
CI 7	Construction	32.10
CI 8	Pre-Commissioning	0.35
CI 9	Commissioning until Performance Tests	0.16
CI 10	Temporary Facilities	5.72
CI 11	Final Documentation	0.25
CI 12	Training of Owner's Operators	0.14
	Total Contract Price	100.00

[a] WF can vary from Project to Project, Owner keeps close control on the allocation and his decision is final and binding. Strategically, WF allocation assumes paramount importance, as progress measurement and payments correspondingly play a very vital role in the cash flow.

TABLE A2.2
Engineering Weight Factor

CI	Description	WF (%)
CI 4.1	Detailed Engineering[a]	57.97
CI 4.2	Procurement Engineering	11.40
CI 4.3	Construction Engineering	13.86
CI 4.4	Pre-commissioning/Commissioning Engineering	3.87
CI 4.5	Start-up and Performance Test Engineering	3.82
CI 4.6	Operation, Maintenance, and Inspection Engineering	5.27
CI 4.7	Interface Engineering	0.49
CI 4.8	Field Engineering	3.32
	Total Engineering CI 4	100.00

[a] The above table shows that Detailed Engineering consumes a greater manpower, which is due to higher number of engineering deliverables under this section.

TABLE A2.3
Procurement and Supply (FOB[a]) Weight Factor

CI	Description	WF (%)
CI 5.1	Vendor/Supplier Assistance	1.24
CI 5.2	Supply of Critical Items	24.59
CI 5.3	Supply of Materials and Equipment for Main Plant	51.15
CI 5.4	Supply of Materials and Equipment for Offsite	15.23
CI 5.5	Consumable and First Fills	1.89
CI 5.6	Capital Spare Parts	4.41
CI 5.7	Operational Spare Parts	0.00
CI 5.8	Additional Equipment to be provided by Contractor	0.93
CI 5.9	Interface Items	0.56
	Total CI 5	100.00

[a] Although FOB is addressed in the section, the supply here means that equipment and materials are ordered by the Contractor to be the supply scope of Vendors. The weight factor allocation shown is between the Contractor and Owner, while that between the Contractor and Vendor could be different. FOB (Free On Board) is defined under Incoterms. The delivery term here is FOB, which is an important aspect in the Purchase Order, which Contractor places on Vendor, as this signifies a battery limit on the Vendor scope. Incoterm definitions are available from the web page in: https://en.wikipedia.org/wiki/Incoterms.

Appendix 3
Monthly Status Report Contents

1. GENERAL
   - 1.1 Monthly Highlights
   - 1.2 Main Meetings Held
   - 1.3 Baseline Schedule Milestones
   - 1.4 Upcoming Holidays and Future Meetings
   - 1.5 Organization Charts and Contact Numbers and Locations
2. HSE, SOCIAL, AND SECURITY
   - 2.1 General Outlook
   - 2.2 Major Activities Done during This Month
   - 2.3 Major Activities Planned for the Next Month
   - 2.4 Outstanding Items
   - 2.5 Status of Social Unrest, HR Issues and Resolutions
   - 2.6 Maintenance Status Report of Residential Camps Civic Amenity and Facilities.
3. PROJECT STATUS
   - 3.1 Project Control
     - 3.1.1 Major Activities Done during This Month
     - 3.1.2 Major Activities Planned for the Next Month
     - 3.1.3 Outstanding Items
     - 3.1.4 Manpower Status (Direct and Indirect with Histograms)
   - 3.2 Engineering
     - 3.2.1 Major Activities Done during This Month
     - 3.2.2 Major Activities Planned for the Next Month
     - 3.2.3 Action Plan for Slippages in Milestones
   - 3.3 Interface
     - 3.3.1 Major Activities Done during This Month
     - 3.3.2 Major Activities Planned for the Next Month
     - 3.3.3 Outstanding Items
   - 3.4 QA/QC
     - 3.4.1 General Outlook
     - 3.4.2 Major Activities Done during This Month
     - 3.4.3 Major Activities Planned for the Next Month
   - 3.5 Authority Interface
     - 3.5.1 General Outlook
     - 3.5.2 Major Activities Done during This Month
     - 3.5.3 Major Activities Planned for the Next Month
     - 3.5.4 Outstanding Items

3.6 Procurement
  3.6.1 General Outlook
  3.6.2 Major Activities Done during This Month
  3.6.3 Major Activities Planned for the Next Month
  3.6.4 Outstanding Items
  3.6.5 Action Plan for Slippages in Milestones
3.7 Transport, Logistics, and Supply
  3.7.1 Major Activities Done during This Month
  3.7.2 Major Activities Planned for the Next Month
  3.7.3 Outstanding Items
3.8 Construction
  3.8.1 Major Activities Done for Main Plant Construction during This Month
  3.8.2 Major Activities for Main Plant Construction Planned for the Next month
  3.8.3 Outstanding Items Related to the Main Plant Construction
  3.8.4 Major Activities Done for Offsites during This Month
  3.8.5 Major Activities for Offsites Construction Planned for the Next Month
  3.8.6 Outstanding Items for Offsite Construction
  3.8.7 Action Plan for Slippages in Milestones
3.9 Temporary Facilities
  3.9.1 Major Activities Done during This Month
  3.9.2 Major Activities Planned for the Next Month
3.10 Option Work Status
3.11 Outstanding ITEMS
3.12 Pre-Commissioning, Commissioning, Performance Tests, and Start Up
  3.12.1 Major Activities Done during This Month
  3.12.2 Major Activities Planned for the Next Month
  3.12.3 Statement of Punch List
  3.12.4 Outstanding Items
3.13 Commissioning, Performance Tests, and Start Up
  3.13.1 Major Activities Done during This Month
  3.13.2 Major Activities Done during the Next Month
  3.13.3 Statement of Punch List
  3.13.4 Outstanding Items
3.14 Maintenance and Inspection Engineering
  3.14.1 Major Activities Done during This Month
  3.14.2 Major Activities Planned for the Next Month
3.15 Following Can Be Attachments to the Monthly Status Report
  3.15.1 Safety Records and Statistics
  3.15.2 Master Document Register Status
  3.15.3 Equipment List updated
  3.15.4 Vendor Drawing Register
  3.15.5 Critical Drawing Progress
  3.15.6 Critical Vendor Drawing List

3.15.7 Purchasing Status Register
3.15.8 Purchase Order Status Report
3.15.9 Material Status Report
3.15.10 Inspection Monitoring Schedule
3.15.11 Inspection Status Report
3.15.12 Spare Part Status Report
3.15.13 Traffic Violation and Improvement Report
3.15.14 Subcontracting Status Report (Award Status, etc.)
3.15.15 Material Delivery Status Report
3.15.16 QA/QC Status Report
3.15.17 Construction Equipment Mobilization
3.15.18 Pre-Com Status Report
3.15.19 Change Order Report
3.15.20 Invoicing Milestone Report
3.15.21 Milestone Status Report
3.15.22 Critical Path Schedule
3.15.23 Detailed Progress Status
3.15.24 Action List from Weekly/Monthly Meetings/No. of Meetings Held
3.15.25 Local Content Progress (Foreign Currency and Local Currency Spent)
3.15.26 Insurance Certificates List (Related to Vendor, Subcontractor and Contractor)
3.15.27 Master Risk Register
3.15.28 Monthly Water Consumption Quantity
3.15.29 Status of Abnormal Weather Report and Work Stoppages
3.15.30 Updated and Detailed Organization Chart with Photo

Appendix 4
Material Requisition Cycle (MR)

Material Requisition (MR) is an engineering deliverable. The following sequence is found in practice. The equipment considered here can fall under Mechanical rotary or static, for example, large columns (like CDU/VDU or CO_2 absorber), reactors, compressors, generators, and high pressure vessels (duplex/double walled). Flow or cycle for other engineering deliverables such a PID, Plant Layout, GA, or Model is not shown in this section.

1. Process department to issue Equipment Data Sheet and this follows update or validation of the data sheet provided in the Front End Engineering Design (FEED).
2. Add Mechanical details into the above Equipment Data Sheet.
3. Issue Equipment Data Sheet/Mechanical Drawing under IFI stage to Owner.
4. Describe the scope of supply, specify licensor requirement, provide a list of applicable specifications, specify the HSE/QA/QC requirements, provide Inspection and Test Plans and Deviation list format, specify key Vendor prints requirement through a Vendor Print Register.
5. Collect input from other disciplines (process, instrument, piping, and civil) as part of the engineering process.
6. Issue Material requisition to Owner.
7. Issue Material requisition after considering Owner comments received from the above step. This issuance can be under Issued for Inquiry, say IFI Rev A. This can be issued to Vendor who is in the approved Vendor list in order to obtain bids/quotes.
8. Receive two quotations from each bidder (Vendor): one is a commercial bid, the other, technical bid in a sealed cover and shortlist the Bidders. One unpriced commercial bid is also required in addition to the above.
9. Technical Bid Evaluation is prepared for shortlisted Bidders after performing some technical clarification and sent to Owner.
10. Owner approves some of the shortlisted Bidders.
11. Contractor conducts commercial negotiation with Bidders and selects a successful bidder and informs Owner.
12. Contractor may have to submit the unpriced together with the final Technical Bid Evaluation Sheet, seeking approval before issuing award notice or letter of intent.
13. Material Requisition issued previously for Inquiry needs to be updated for issuing this for Purchase. For this, document such as Basis of Design, PFD, HMB, MSD, Licensor specifications, PID, Plot Plan, Hydraulic (pump) calculation Notes, Design calculations on thermal rating, utility

consumption, strength calculations, etc. are to be done. In addition to this, the documents mentioned under step 4 are also to be updated after clarifications with the selected Bidder.

14. Equipment Data Sheet updated in the above step may have to be submitted to Owner before issuing Material Requisition for Purchase.

Note: The steps and flow are made until the PO is issued. Thereafter, if a change takes place, such changes are captured in revisions of the MR so issued for PO. Some Bidders although qualify technically may find disqualified if they are not in a position to provide bank guarantees from the banks approved by the Owner. The Owner may also not approve if the financial profile of the Bidder is not sound. The life cycle cost of each shortlisted bidder is to be obtained and analyzed if capital cost is less than future operating/maintenance/disposable cost.

Appendix 5
FEED Deliverable List

Following is the list of deliverables in brief, which may vary from Project to Project and Owner to Owner.

- Plant Layout
- Unit Layout
- Process Block Flow Diagram
- Design Basis
- Duty Specifications
- Cause and Effect Diagrams
- Process Description
- Process Flow Diagram
- P&IDs
- Heat and Material Balance
- Material Selection
- Equipment List
- Equipment Data Sheets
- Instrumentation Data Sheets, I/O Summaries
- Summaries related to Flare load, Electrical load, utility consumption, catalyst and Chemicals, and effluents
- Engineering Flow Diagrams
- Interlock Description, Cause and Effect, Control Narratives
- HAZOP/SIL report
- 3D Model
- Architecture for control systems
- Philosophies (Fire Protection, Safety Concept), Project Procedures, Job Specifications and Engineering specifications, and List of standards and codes may or may not be part of FEED; it is the choice of the Owner whether to include these under FEED or otherwise.

Appendix 6

Process Block Flow Diagram

See Figure A6.1.

TYPICAL PROCESS BLOCK FLOW DIAGRAM FOR UPSTREAM GAS PROCESSING UNIT

Process Stream Path and Description: The process fluid from Wells runs through Cluster, Production Manifolds, Slug Catcher. Fluid splits into gas and liquid at downstream of slug Catcher. Then the gas stream runs through Mercury Removal unit, Compression unit, Acid Gas removal unit, Dehydration Unit and HC Dew Point Unit and finaly into Fiscal Metering Unit before being sent into export gas pipeline. The liquid stream from Slug Catcher runs into LP Seperator from which gas stream goes into Fuel Gas unit before being used in GTG for firing, while liquid stream into Condensate Stabilization, Condensate Metering, and finally Condensate storage. At Condensate storage condensate will be loaded into the trucks before being sent to consumers distribution network. Thermal oxidizer is to oxidize the off gas from Acid Gas Recovery Unit before letting the gas into atmosphere. This kind of oxidation to neutralize polluntants depends on environmental specifications of each country. PWT(Produced Water Treatment) unit is to treat the water coming from various process units and helps recyling the water.

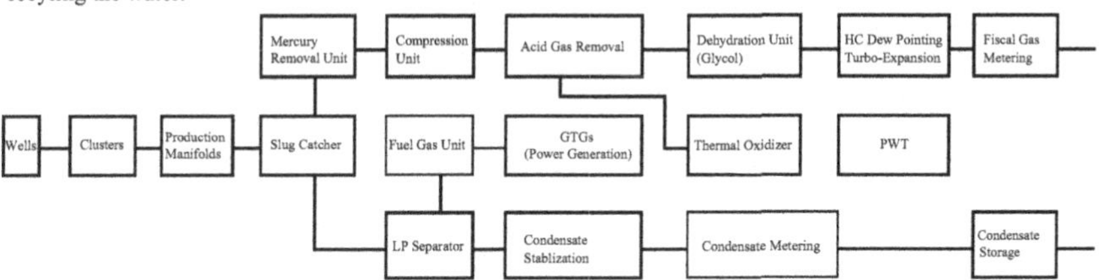

FIGURE A6.1 Typical process block flow diagram for an upstream gas processing unit.

The PWT unit collects water from MP Compression, Acid Gas Removal Unit, and LP Separator and treats it. Well and Clusters are considered as part of Off-Site. Flowlines carry gas from wells to Clusters, and Trunklines carry gas from Clusters to Production Manifolds.

Appendix 7

Document Categories

1. Document Categorization

The document (drawings/procedures/plans/job specifications/etc.) which a Contractor makes can fall into several categories. Let us assume it has three categories.

- Category 1
- Category 2
- Category 3

Category 1 Document: Category 1 Documents are those that the Owner seeks to closely monitor, review, and approve. Category 1 documents may include but not limited to P&ID, Heat Mass Balance, Layouts appearing under the stages of Drawings-issued-for-Design and Drawings-issued-for-Construction, etc. Without the approval of Category 1 document by Owner, Contractor cannot take up fabrication/manufacture or construction works pertinent thereto, and hence a stringent quality control is consequently assured. As these documents generally table in the critical activities list, approval not obtained on time affects the Critical Path of the Project.

Category 2 Document: Category 2 Documents are those that Owner seeks to review and comment. The delay or non-approval from Owner does not prevent Contractor from carrying out the next step of execution. However, Owner may expect his comments implemented in later stages.

Category 3 Document: Category 3 Documents are those that Owner receives from Contractor for "Information Only."

In the following, 11 tables show documents related to various disciplines. For some documents, the category number is indicated and the rest is left blank.

In Contract documents, these are generally provided and assigned with category numbers. However, the Documentation Procedure is made during the execution phase to finalize the extent of documentation to be submitted to Owner and category thereof between Owner and Contractor. A Class of Approval should be described in the Procedure. The turnaround time for review cycle of each category of documents is addressed in the document transmittal sheet with a precise summary of past review date and duration taken. Using one common transmittal sheet for a bulk of documents creates difficulty in the Document Management System and hence individual transmittal is recommended.

2. Documents of Common Nature (Table A7.1)
3. Process Engineering Documents (Table A7.2)
4. Piping Documents (Table A7.3)
5. Pipelines Documents (Table A7.4)
6. Instrument Engineering Documents (Table A7.5)
7. Electrical Engineering Documents (Table A7.6)
8. Telecom Engineering Documents

TABLE A7.1
Documents of Common Nature

Document Name	Category
Procedures developed by Contractor under each discipline	2
90 Days Starter Schedule, Level 3 Detailed Schedule	2
Work Breakdown Structure (WBS)	2
Vendor List, Subcontractor List	1
Requisition Index	
Equipment List	
Manufacturers Record Book	
Operation Manual	
Spare Parts Index	

TABLE A7.2
Process Engineering Documents

Document Name	Category
Process flow diagrams	1
Heat and mass balance	1
Material construction diagram	1
P&ID	1
Piping line list	1
HAZOP reports	
Lubricant list	
MSDS	1
Utilities consumption as designed	
Pressure drop profile	1
Cause and effect diagrams	1

TABLE A7.3
Piping Documents

Document Name	Category
General arrangement piping drawings	2
Calculations piping	1
Details of expansion joints	3
Details piping support	3
Piping standards (including sketches)	2
Critical lines (isometrics)	2
Tie-ins (packages if existing piping to be tied)	2

TABLE A7.4
Pipeline Documents

Document Name	Category
Flow assurance study	
Pipeline route survey	
General arrangement	
Pipeline alignment drawings	
Trench details	
Pressure drop calculations	
Tie-ins	
Drawings for crossings	

If this part of work is subcontracted to approved Subcontractors, then Contractor and Owner jointly decide on what documents to be submitted to Owner.

9. Civil Engineering Documents (Table A7.7)

10. MR (Material Requisitions)

    For every equipment, there is one MR. For Bulk materials, MR can be individualized according to the category; piping, structural, electrical, and instrument have bulk materials (Table A7.8).

11. Vendor Prints/etc.

    Prior to Contractor issuing Purchase Order to Vendor, Contractor finalizes a list of Key Vendor prints that would require submission later to Owner through Contractor. Technical Bid Evaluation summary sheets are submitted to Owner for comments and approval depending on the criticality and importance of the equipment and instruments and other items.

TABLE A7.5
Instrument Engineering Documents

Document Name	Category
Index of instruments (instrument list)	1
Data Sheets of instruments	1
Layout in the control room	1
Layout of instrument panels situated in the control room or at field	
Fire and gas detection system layout	1
Layout of fire and gas panels	1
Single line diagrams (instrument power supply system)	1
Wiring diagrams	2
Diagrams (for functional logic) for safeguarding and sequence	1
Diagrams typical (earthing)	2
Route diagram for the main cable	1
Loop diagrams typical for critical items especially for process measurements)	1
Trouble shooting loop diagram	2
Hook up diagrams typical	2
Design calculation for orifices, relief valves, control valves, etc.	
Set points for alarm and trip	1
Level sketches	1

TABLE A7.6
Electrical Engineering Documents

Document Name	Category
Single line diagrams	1
Hazardous area classification layout (in coordination with the piping discipline)	1
Earthing philosophy and typical details	2
Route drawing (main cable)	2
Electrical equipment rating summary	
Load balancing study and reports	1
Voltage drop calculations	1
Typical drawing or standard sketches for installation	
Fault and load flow analysis	2
Relay co-ordination	2
System protection	2
Electrical drawings	2

12. Construction/Pre-commissioning/Commissioning Documents (Table A7.9)
13. HSE Documents (Table A7.10)
14. Quality Control and Assurance Documents (Table A7.11)

TABLE A7.7
Civil Engineering Documents

Document Name	Category
Drawing for earth work	
Layout drawing for piling (if any piling work is involved)	
Layout drawing for any buildings such as control room including calculations	
Layout drawing for trenches that carry electrical and instrument cables	
Layout for paving	3
Layout for any buried/underground services	
Schematic drawing for drainage	1
General arrangement drawing for structural steel works	
Foundation drawings for equipment and structures	
Architecture drawings	1
Building: layout and GA and calculations for services	1

TABLE A7.8
MR(Material Requistions)

Document Name	Category
MR for equipment/instruments/bulk materials/etc.	1
MR for equipment/instruments/ bulk materials/etc.	2

TABLE A7.9
Constructio/Pre-commissioning/Commissioning Documents

Document Name	Category
Plot plan and construction laydown areas (temporary facility).	2
Hauling and erection of heavy and critical equipment (plan/procedure/calculations)	2
WPS, PQR, WPQT	1
Method statements	3
Procedures for pre-commissioning	2
Procedure and checklist sheets for software like ICAPS	2
Procedures for commissioning	1

TABLE A7.10
HSE Documents

Document Name	Category
Evacuation plans	1
Procedures for scaffolding check	2
Procedure for induction and gate pass/id badges	2
Procedure for housekeeping and environmental check	2
Procedure for wearing safety belt/helmet and shoes and harness	2
Procedure for vehicle entry and parking	3

TABLE A7.11
Quality Control and Assurance Documents

Document Name	Category
Audit plans	2
Procedure for issuing and disposing NCR	2
Procedure for tagging NCR issued jobs	2

Appendix 8
Claims and Liens Release Letter

To
[Contractor's name and address]

We, for and in consideration of our receipt of the sum of the amount ___________ from you, representing the final payment under the Purchase Order/Subcontract between you and us that has become in full force and effective as of the effective date including all amendments thereof, hereby release and forever discharge you and Owner from all claims and demands whatsoever in any manner arising out of, or related to, labour performed or materials and equipment furnished by us under the Purchase Order/Subcontract in connection with, or incidental to, the construction of the Project or title for the Owner.

In consideration of, and for the purpose of inducing you to make the aforesaid final payment, we hereby warrant that we shall indemnify and hold harmless you and the Owner from all demands, liens, and claims of any nature including tort claims in any manner arising from or related to the Purchase Order/Subcontract.

We understand that the foregoing shall not relieve us of our obligations under the provisions of the Purchase Order/Subcontract, which by their nature survive the Completion of Work including, without limitation, warranties, guarantees, and indemnities.

In Witness Whereof, we have caused this instrument to be executed by our duly authorized officer this ___________ day of year ___________

Signed ___________

Name of the Vendor/Subcontractor

Appendix 9

Prolongation and Disruption Cost: Hudson Formula

A9.1 PROLONGATION COST

The prolongation cost can be assessed by using Hudson's formula with the assumption as below.

In this Project, a project delay of 10 months is assumed to have occurred due to acts and omissions of Owner. The 10 month delay mentioned is arrived at after removing the delay corresponding to the concurrent delays.

H Head office overhead %, to be used from Contract Price Breakdown that may show this, else Contractor should compute the actual expenses from SAP on this category and Owner will check this. Let us assume this 14% (where 6% overhead and 8% for profit) for the purpose of computation.

C Contract Price, 1000 million USD

t Period of Delay, 10 months

T Contract Period, 40 months

Prolongation cost = $(H/100) \times (t/T) \times C = (14/100) \times (10/40) \times 1000 = 35$ million USD

[For detailed information, reference is made to the article:
A Formula For Success, by John B. Molloy, James R. Knowles (Hong Kong) Limited]

A9.2 DISRUPTION COST

Let us assume Contractor suffers disruption cost from the acts or omissions of Owner in delaying the approval of critical engineering deliverables PID. Since the PID is in critical path, the delay results in overall delay of the Project when time event analysis is done on the P6 Primavera Level III schedule. A Contractor by keeping records of additional engineering man-hours spent can have the opportunity of getting the claim approved.

D Disruption cost
M total man-hours additionally consumed on the delay of PID approval
U unit rate of engineers, averaged, e.g., 200 USD/hour as allowed in Unit Rates of Contract

$D = M \times U$, or
$D = M1 \times U1 + M2 \times U2 + M3 \times U3$.... This is a more accurate method, where M1 and U1 may stand for Senior Process Engineer, M2 and U2 for Piping Engineer, and M3 and U3 for Instrument engineer.

Note: The cost allowed is only for the idle man-power. No allowance can be considered for profit and overhead. While cost is captured in Disruption Cost, the delay in the time is captured in Time extension and its cost is captured in Prolongation Cost.

Appendix 10
A Typical List of Procedures and Plans

1. Management Procedures

- Project Management Procedure;
- Project Co-Ordination Procedure including correspondence numbering;
- Numbering Procedure;
- Change/Variations Procedure;
- Invoicing Procedure;
- Scheduling Procedure;
- Cost Control Procedure;
- Document Filing Procedure;
- Document Management Procedure;
- Progress Measurement Procedure including Work Breakdown structure (between Contractor and Owner). Excel sheet showing how progress is to be calculated;
- Guidelines for Payment break up for Vendors and Subcontractors and qualification;
- Progress Reporting Procedure;
- Authority Approvals Matrix and Procedure;
- Punch List Procedure;
- As-Built Documentation Procedure;
- Risk Assessment Procedure;
- Project Milestones and Payment Milestones summary;
- Organization Chart and histograms for direct and indirect manpower: monthly update is needed.

2. Engineering Procedures

- Engineering Procedure;
- Design HSE Procedure;
- Co-ordination Procedure between Licensors and Contractor;
- List of Codes and Specifications;
- Formats for Requisition and Technical Bid Evaluations, Deviations;
- Compile a list of International Codes and Standards with revisions;
- Procedure for studying piping flexibilities.

3. Procurement Procedures

- Project Procurement Plan;
- Equipment and Material List (inputs are from Engineering disciplines);
- General Terms and Conditions for PO to Vendors;
- Expediting Procedure;
- Shipping and Logistic Procedure;
- Customs Clearance Procedure;
- Packing and Marking Procedure;
- Vendor Documentation Procedure;
- Procedure for Spare Parts Price List;
- Format for Procurement and Delivery Status reports. Provide a separate annexure to specify what items are not delivered per each PO and Requisition.

4. Construction Procedures

- Construction Co-Ordination Procedure;
- Sub-Contracting Plan;
- Method Statements for each important item of work per discipline;
- Procedures for Painting;
- Procedure for insulation;
- Procedure for Cathodic Protection tests;
- Procedure for grounding;
- Procedures for equipment installation as per category;
- Procedures for Piping work Installation;
- Plans for shop fabrication;
- Procedure for storage;
- Procedure for Field Material Management.

5. Pre-Commissioning Procedures

- Safety Plan for Pre-Commissioning;
- Pre-Commissioning Plan;
- Pre-Commissioning Sequence of sub-systems and systems.

6. Commissioning, Start Up and Performance Test Procedures

- Commissioning Plan;
- Commissioning Dossier;
- Commissioning Sequence;
- Pre-Start Up Safety Plan;
- Procedure for Performance Test Runs;
- Procedure for loading chemicals, oils, etc.;
- Energization Procedure;
- Laboratory Test Procedure.

7. Quality Procedures

- Project Quality Plan;
- Quality Audit Procedure;
- Inspect and Test Procedure (for Vendor shop and Site separately);
- ITP for each category of work per discipline;
- Procedures for NDT (at shop and site);
- Procedure for concrete testing at Site;
- Non-Conformance Report Procedure including correction action to be followed;
- Procedure for preparing WPS, PQR, and WQT;
- Pressure test (hydro and pneumatic) Procedure for piping;
- Calibration Procedures for various instruments and equipment (Permanent and temporary).

8. HSE Procedures

- Health, Safety and Environment Plan for Site;
- Emergency Evacuation Plan;
- MEDEVAC Plan (to be from qualified agency);
- HSE Compliance policy and enforcement plan;
- Job Hazard Analysis for each work item having risk in execution prior to execution;
- Procedure for Scaffolding;
- Procedure for construction equipment and vehicles use;
- PPE guidelines and policy;
- Safety Lock out/Tag Out Procedure (for each important activity that involves the use of electricity);
- Procedure for safety Induction and issuance of identity pass;
- Procedure for barricading and traffic movement inside site;
- Procedure for using electrical switch boards and connections for construction purposes and temporary power usage.

Appendix 11
Total Plant Breakdown into Areas

A Plant/Project may have both Main Plant and Off-Site Facilities in the oil and gas upstream sector. The following areas are typical for the Main Plant. These areas when put together represent geographically the entire Plant/Project.

For planning and execution of activities of a Project in Level 3, area-wise representation is popular and suitable. In Level 3, some activities fall under General/Common without representing any area in particular, while most fall under respective areas.

1. Reception Area;
2. AGRU and Process Area;
3. Gas Export and Storage Area;
4. Utilities Facilities (which may also spread across Process Areas);
5. Building Area;
6. Flare Zone, and;
7. Pipe Rack area.

Bibliography

AUTHORS

Martin Hunter and Alan Redfern, *Law and Practice of International Arbitration, fourth edition by London Sweet & Maxwell 2004, ISBN 9780421862401,0421862408,* https://books.google.co.in/books/about/Law_and_Practice_of_International_Commer.html?id=9mBqDaSB-ZwC&redir_esc=y.

S. Rahman and A. Dana Vanier, *Life cycle cost analysis as a decision support tool for managing municipal infrastructure, published by National Research Council Canada, May 2-9, 2004 in CIB 2004 Triennial Congress, Toronto, Ontario.* www.irbnet.de/daten/iconda/CIB9737.pdf. http://energy.gov/sites/prod/files/2014/05/f16/pumplcc_1001.pdf

Hydraulic Institute, Europump, and US Department of Energy's office of Industrial Technologies(OIT). *Pump Life Cycle Cost: A Guide to LCC Analysis for Pumping Systems by*

John B. Molloy and James R. Knowles, *A Formula for Success.* https://www.hkis.org.hk/ufiles/dis03.pdf

WEBSITES

Wikipedia. Amine Gas Treating. https://en.wikipedia.org/wiki/Amine_gas_treating.

Legal Information Institute. Ejusdem Generis. https://www.law.cornell.edu/wex/ejusdem_generis.

Index

Printed in the United States
By Bookmasters